KUMON MATH WORKBOOKS

Grades 0-8

Algebra
Workbook I

Table of Contents

KUM☺N

Fractions Review

Date   /   /

Name

Score

/100

■ The Answer Key is on page 88.

1 **Calculate.**

5 points per question

(1) $0.5 + \dfrac{1}{9} =$

(6) $1\dfrac{3}{8} - 1.3 =$

 Rewrite decimals as fractions to calculate.

(2) $0.6 - \dfrac{1}{4} =$

(7) $0.75 - 0.25 + \dfrac{1}{2} =$

(3) $\dfrac{2}{5} + 0.75 =$

(8) $4.5 + \dfrac{1}{4} - 1.125 =$

(4) $1.75 - \dfrac{1}{6} =$

(9) $3\dfrac{1}{2} - 0.375 + 5\dfrac{1}{6} =$

(5) $\dfrac{3}{10} + 2.125 =$

(10) $9.25 - \dfrac{1}{2} + \dfrac{3}{8} =$

2 **Calculate.**

5 points per question

(1) $0.125 \times \frac{1}{4} =$

(6) $2.25 \div 1.5 =$

(2) $0.75 \div \frac{6}{11} =$

(7) $\frac{1}{3} \times 2.7 \times \frac{4}{5} =$

 Remember to reduce as you calculate.

(3) $\frac{2}{3} \times 0.6 =$

(8) $\frac{3}{10} \div 1.4 \div \frac{3}{4} =$

(4) $\frac{1}{2} \times 3.5 =$

(9) $2.4 \times \frac{5}{6} \div 10 =$

(5) $0.375 \times 6 =$

(10) $1.8 \div 0.375 \times 1.25 =$

You're off to a great start!

3

2 Exponents Review

Date / /

Name

Level ☆ ....

Score

/100

■ The Answer Key is on page 88.

1 **Calculate.**

5 points per question

(1) $2^5 \times \dfrac{1}{2^2} =$

(6) $2^2 \times 2^1 \times 2^2 = 2^{\square}$

$=$

> When multiplying exponents with the same base, add the exponents.

(2) $\left(\dfrac{3}{10}\right)^2 \times 2^4 =$

(7) $\left(\dfrac{2}{3}\right)^3 \times \left(\dfrac{2}{3}\right)^1 =$

(3) $\left(\dfrac{4}{9}\right)^2 \times \left(\dfrac{3}{8}\right)^2 =$

(8) $\dfrac{3^2}{4^1} \times \dfrac{3^0}{4^2} =$

> 2^1 is the same as 2.

(4) $\left(1\dfrac{1}{2}\right)^3 \times \left(\dfrac{8}{15}\right)^2 =$

(9) $\dfrac{2^3}{3^1} \times \dfrac{2^2}{3^1} \times \dfrac{2^1}{3^2} =$

> Any number raised to the 0 power equals 1.

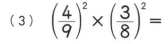

(5) $\left(2\dfrac{1}{4}\right)^4 \times \left(1\dfrac{1}{3}\right)^3 =$

(10) $\dfrac{2^3}{5^1} \times \dfrac{3^0}{4^1} \times \dfrac{2^2}{5^0} \times \dfrac{3^2}{4^1} =$

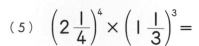

4   © *Kumon Publishing Co., Ltd.*

2 **Calculate.**

(1) $2^6 \div 2^2 = 2^{\square} =$

(6) $\left(\dfrac{5}{2}\right)^2 \div (15)^2 =$

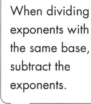

When dividing exponents with the same base, subtract the exponents.

(2) $3^{10} \div 3^4 \div 3^1 =$

(7) $\left(2\dfrac{1}{4}\right)^3 \div \left(3\dfrac{3}{4}\right)^3 =$

(3) $\left(\dfrac{2}{3}\right)^5 \div \left(\dfrac{2}{3}\right)^1 =$

(8) $\left(\dfrac{1}{2}\right)^2 \div \left(\dfrac{1}{2}\right)^1 \times \left(\dfrac{1}{2}\right)^3 =$

(4) $10^3 \div 2^3 =$

(9) $\left(\dfrac{3}{5}\right)^3 \times 6^3 \div \left(\dfrac{9}{10}\right)^3 =$

(5) $9^3 \div 12^3 =$

(10) $\left(1\dfrac{1}{3}\right)^2 \div \left(1\dfrac{2}{3}\right)^2 \times \left(1\dfrac{7}{8}\right)^2$

$=$

Your skills are strong!

3

Order of Operations Review

Level ☆ ....

Score

/100

Date / /

Name

■ The Answer Key is on page 88.

1 **Calculate.**

5 points per question

> **Don't forget!**
>
> According to the order of operations,
> - **calculate exponents and numbers in parentheses and brackets first**
> - perform multiplication and division before addition and subtraction
> - calculate from left to right

(1) $1 + 6 - 4 \div 8 =$

(5) $\dfrac{1}{2} \div \left(\dfrac{1}{3} \times 2 \right) + 7 =$

(2) $1 + (6 - 4) \div 8 =$

(6) $\left(\dfrac{1}{2} \div \dfrac{1}{3} \right) \times (2 + 7) =$

(3) $[(1 + 6) - 4] \div 8 =$

(7) $\dfrac{2}{3} + 1\dfrac{1}{2} \div \dfrac{1}{4} \times 1\dfrac{1}{3} =$

(4) $\dfrac{1}{2} \div \dfrac{1}{3} \times 2 + 7 =$

(8) $\dfrac{2}{3} + 1\dfrac{1}{2} \div \left(\dfrac{1}{4} \times 1\dfrac{1}{3} \right) =$

② Calculate.

(1) $2 \div 6 + \left(\dfrac{1}{2}\right)^2 + 1 =$

(4) $3\dfrac{1}{3} + 2^2 - 9 \div 4 \times 2 =$

(2) $6 \div \left(2 + \dfrac{1}{2}\right)^2 + 1 =$

(5) $\left(1\dfrac{1}{3} + 2\right)^2 - 9 \div (4 \times 2)$

$=$

(3) $6 \div \left[\left(2 + \dfrac{1}{2}\right)^2 + 1\right] =$

(6) $\left[\left(1\dfrac{1}{3} + 2\right)^2 - 9\right] \div (4 \times 2)$

$=$

Nicely done!

Negative Numbers Review

Level

☆

Score

/100

Date / /

Name

■ The Answer Key is on page 88.

1 **Calculate.**

2 points per question

(1) $3-7=$

(2) $-2+8=$

(3) $6-(-7)=$

(4) $\dfrac{1}{4}+(-1)=$

(5) $-\dfrac{2}{5}+7=$

(6) $-2-\left(-1\dfrac{1}{3}\right)=$

(7) $-2\dfrac{2}{3}-\dfrac{1}{4}=$

(8) $-3\dfrac{2}{5}-\left(-1\dfrac{1}{4}\right)=$

(9) $1-2\dfrac{1}{2}=$

(10) $-1-2\dfrac{1}{2}=$

(11) $-1-\left(-2\dfrac{1}{2}\right)=$

(12) $7\dfrac{1}{3}-2\dfrac{1}{2}=$

(13) $-7\dfrac{1}{3}+\left(-2\dfrac{1}{2}\right)=$

(14) $-7\dfrac{1}{3}-\left(-2\dfrac{1}{2}\right)=$

② Calculate.

6 points per question

(1) $6 \times (-2) =$

(2) $-4 \times (-8) =$

(3) $-\dfrac{4}{5} \times \dfrac{3}{8} =$

(4) $1\dfrac{4}{5} \times \left(-3\dfrac{1}{3}\right) =$

(5) $18 \div (-3) =$

(6) $(-24) \div (-12) =$

(7) $-\dfrac{1}{4} \div 10 =$

(8) $\dfrac{1}{2} \div \left(-\dfrac{3}{8}\right) =$

(9) $9 \times \left(-1\dfrac{2}{3}\right) \div \left(-3\dfrac{1}{3}\right) =$

(10) $-2 \div \left(-2\dfrac{2}{5}\right) \times \left(-1\dfrac{4}{5}\right) =$

(11) $-3\dfrac{1}{5} \div \dfrac{8}{15} \div 9 \times \left(-1\dfrac{1}{8}\right)$

$=$

(12) $6\dfrac{2}{3} \times \left(-2\dfrac{1}{4}\right) \div \left(-3\dfrac{1}{3}\right) \div \left(-1\dfrac{2}{3}\right)$

$=$

When multiplying or dividing with negative numbers, count the number of negative signs and determine the sign of the answer first. Then calculate.

Positively wonderful work!

Level
☆

Score

/100

Date / /

Name

■ The Answer Key is on page 88.

1 **Calculate.**

5 points per question

(1) $(-2)^5 =$

(2) $(-3)^3 \times (-2)^2 =$

(3) $6^2 \times (-3)^0 =$

(4) $(-8) \times \left(-\dfrac{1}{4}\right)^3 =$

(5) $-\left(-1\dfrac{1}{3}\right)^4 \times \left(-1\dfrac{1}{8}\right)^2 =$

(6) $(-2)^4 \div (-4)^3 =$

(7) $-2^4 \div (-4)^3 =$

(8) $3^3 \div \left(-\dfrac{1}{2}\right)^2 =$

(9) $\left(-\dfrac{3}{5}\right)^2 \div \left(-\dfrac{9}{10}\right)^3 =$

(10) $\left(-2\dfrac{1}{2}\right)^3 \div \left(1\dfrac{7}{8}\right) =$

Pay close attention to the placement of parentheses.

© Kumon Publishing Co., Ltd.

2 **Calculate.**

5 points per question

(1) $\dfrac{-\dfrac{1}{2}}{\dfrac{3}{4}} =$

(2) $\dfrac{\dfrac{3}{7}}{-\dfrac{6}{11}} =$

(3) $\dfrac{8}{\dfrac{1}{3} - \dfrac{1}{2}} =$

(4) $\dfrac{-\dfrac{1}{4} - \dfrac{1}{8}}{5} =$

(5) $\dfrac{\dfrac{1}{4} - \dfrac{3}{5}}{2\dfrac{1}{4} - 1\dfrac{1}{2}} =$

(6) $(-1)^6 \times (-3)^2 \div (-2)^3$

$=$

(7) $3^4 \div (-2)^5 \times 6^0 \div (-3)^2$

$=$

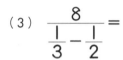

Remember to follow the order of operations.

(8) $(-2) - (-4)^3 \div (-3)^2 + (-5)^2$

$=$

(9) $\left(-\dfrac{2}{5}\right)^2 \times (2 + 6 \times 3) \div (-2)^3$

$=$

(10) $\left(-3\dfrac{2}{3}\right)^0 \div \left(-\dfrac{3}{4} - \dfrac{1}{4} \times 2\right) + (-12 \div 10)^2$

$=$

If you have difficulty with this workbook, please try *Pre-Algebra Workbook II*.

Great job!

11

Date / /

Name

■ The Answer Key is on page 88.

1 **Determine the value of each expression when** $x = 3$.

5 points per question

(1) $-4x =$

(2) $\dfrac{5}{6}x =$

(3) $x - 9 =$

(4) $2x + 4 =$

(5) $-5x - 6 =$

(6) $\dfrac{1}{2}x - \dfrac{9}{2} =$

(7) $8 \div \dfrac{1}{x} =$

(8) $6 - \dfrac{3}{2x} =$

(9) $8x - \dfrac{5}{2x} =$

(10) $-\dfrac{3}{2}x + \dfrac{x}{4} =$

2 **Determine the value of each expression when** $s = \dfrac{2}{3}$**.**

(1) $5s =$

(6) $\dfrac{2-s}{s} =$

(2) $2s - \dfrac{1}{4} =$

(7) $\dfrac{4-s}{3s+2} =$

(3) $6 - \dfrac{1}{4}s =$

(8) $\dfrac{s}{2} - 3 - \dfrac{1}{s} =$

(4) $\dfrac{s}{4} =$

(9) $-\dfrac{7}{10}s \div 2s =$

(5) $-\dfrac{6}{s} =$

(10) $\dfrac{s + \dfrac{1}{2}}{1 - s} =$

Keep up the good effort!

Level
☆

Score

/100

■ The Answer Key is on page 89.

1 **Determine the value of each expression when** $j = -2$.

5 points per question

(1) $j^5 =$

(6) $j^2 - 9 =$

(2) $-j^3 =$

(7) $(j+3)(j-3) =$

(3) $3 - j^2 =$

(8) $j^4 - j^3 =$

(4) $\dfrac{3}{j^2} =$

(9) $j^5 - j^2 =$

(5) $\dfrac{-9-j}{j^2+1} =$

(10) $j^2(j^3 - 1) =$

 Pay close attention to the placement of negative signs.

2 **Determine the value of each expression when** $z = -\dfrac{1}{4}$ **.**

5 points per question

(1) $3\dfrac{1}{5} + z =$

(6) $z^3 - z =$

(2) $4z - 6 =$

(7) $\dfrac{2}{z} + \dfrac{3}{z^2} =$

(3) $\dfrac{z}{6} =$

(8) $\dfrac{z^2 + 1}{z} - z =$

(4) $\dfrac{10}{z + 1} =$

(9) $2z^2 + 5z + 2 =$

(5) $z^2 + z =$

(10) $(2z + 1)(z + 2) =$

Smart thinking!

15

Values of Algebraic Expressions Review

Date   /   /

Name

■ The Answer Key is on page 89.

1   **Determine the value of each expression when** $x = 1$ **and** $y = -2$.   5 points per question

(1) $2x + 3y =$

(2) $y - \dfrac{1}{2}x =$

(3) $-\dfrac{3}{4}x + \dfrac{1}{3}y =$

(4) $-\dfrac{9}{2}x - 2y =$

(5) $-\dfrac{3}{2}y + \dfrac{9}{4}x =$

(6) $4xy + \dfrac{1}{x} =$

(7) $\dfrac{y}{3x} - x =$

(8) $9x^2 - 4y^2 =$

2 **Determine the value of each expression when** $a = -1$, $b = 2$, 10 points per question
and $c = -\dfrac{1}{2}$.

(1) $\dfrac{a}{b} + \dfrac{b}{c} + \dfrac{c}{a} =$

(4) $-(-a)^2 + (-b)^2 + c^3 =$

(2) $ab + bc =$

(5) $(b^2 - c^2) - (a^2 - c^2) =$

(3) $b(a + c) =$

(6) $(b + c)(b - c) - (a + c)(a - c)$

$=$

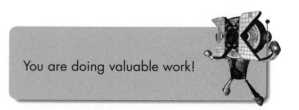

You are doing valuable work!

Values of Algebraic Expressions Review

Date ／ ／

Name

Score ╱ 100

■ The Answer Key is on page 89.

1 Determine the value of each expression when $a = -3$, $b = 2$, $c = -1$, and $d = \dfrac{3}{2}$.

5 points per question

(1) $abcd =$

(2) $\dfrac{a+b}{c-d} =$

(3) $\dfrac{a}{b} - \dfrac{c}{d} =$

(4) $\dfrac{ad-bc}{bd} =$

(5) $b^0 - c + d^2 - a^3 =$

(6) $(c - ad)(a^2 + ab + b^2) =$

(7) $[(a \div b) \div (c \times d)]^3 =$

(8) $a \div [b \div (c \times d)]^2 =$

2 **Use the given values of w, x, y, and z to determine the value of the expression $\dfrac{x^2y}{z} - \dfrac{z}{w-x}$.**

10 points per question

(1) $w=3$, $x=0$, $y=1$, $z=2$

$$\frac{x^2y}{z} - \frac{z}{w-x} =$$

(4) $w=1$, $x=\dfrac{1}{2}$, $y=-\dfrac{1}{3}$, $z=-\dfrac{1}{4}$

$$\frac{x^2y}{z} - \frac{z}{w-x} =$$

(2) $w=-4$, $x=-3$, $y=-1$, $z=-2$

$$\frac{x^2y}{z} - \frac{z}{w-x} =$$

(5) $w=\dfrac{1}{2}$, $x=\dfrac{2}{3}$, $y=\dfrac{1}{4}$, $z=-\dfrac{3}{2}$

$$\frac{x^2y}{z} - \frac{z}{w-x} =$$

(3) $w=2$, $x=-1$, $y=1$, $z=-2$

$$\frac{x^2y}{z} - \frac{z}{w-x} =$$

(6) $w=-2\dfrac{1}{3}$, $x=-1\dfrac{1}{3}$, $y=\dfrac{1}{2}$, $z=-2$

$$\frac{x^2y}{z} - \frac{z}{w-x} =$$

Bravo!

Simplifying Algebraic Expressions

Date / /

Name

Score /100

■ The Answer Key is on page 89.

1 **Simplify each expression.**

4 points per question

Examples

$3x + 4x = 7x$ $8a - 2a = 6a$

(1) $2x + 3x = 5\square$

(2) $6y + 9y =$

(3) $8d + 0 =$

(4) $x + 3x =$

(5) $2z + 5z + z =$

(6) $9b - 4b =$

(7) $10h - 7h =$

(8) $5x - 4x =$

> Always write x or $-x$, not $1x$ or $-1x$.

(9) $12y - 5y - 4y =$

(10) $15w - 6w + 13w =$

— Don't forget! —

A variable is a letter or symbol that represents an unknown amount. Variables can be any symbol or letter, such as x, y, a, $\Diamond$, etc.

2 **Simplify each expression.**

Example

$4a - a = 3a$

It's okay to leave your answers as improper fractions when there is a variable.

(1) $5x + 9x =$

(2) $7y - 4y =$

(3) $5a - (-4a) = 5a + \boxed{} =$

(4) $-3h + (-10h) =$

(5) $6a - (-a) =$

(6) $5j - 0 =$

(7) $0 - 5j =$

(8) $\dfrac{5}{7}w - \dfrac{3}{7}w =$

(9) $-\dfrac{2}{9}z + \dfrac{7}{9}z =$

(10) $\dfrac{1}{4}f + \left(-\dfrac{3}{4}f\right) =$

(11) $2x + \dfrac{1}{3}x = \dfrac{\boxed{}}{3}x + \dfrac{1}{3}x = \dfrac{\boxed{}}{3}x$

(12) $\dfrac{3}{4}y + 2y =$

(13) $3b - \dfrac{2}{3}b =$

(14) $-\dfrac{1}{4}m - \left(-\dfrac{5}{4}m\right) =$

(15) $7xy - 4xy = \boxed{}xy$

Don't forget to reduce your answers!

Great start on a tough topic!

Simplifying Algebraic Expressions

Date / /

Name

Score

/100

■ The Answer Key is on page 89.

1 **Simplify each expression.**

4 points per question

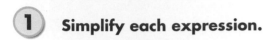

> **Don't forget!**
> Combine like terms.   **Example** $3x + 4 + 2x + 7 = 5x + 11$

(1) $2x + 4x + 5 = \boxed{}x + 5$

(8) $2a + 4b + 3b - 6a = \boxed{}a + \boxed{}b$

(2) $10y - 3 + 4y =$

(9) $-\dfrac{5}{3}b + 2a + \dfrac{1}{3}b - a =$

(3) $-6 + \dfrac{7}{3}h - \dfrac{1}{3}h =$

(10) $4w - (-w) - y + 2y =$

(4) $\dfrac{3}{4}bc + \dfrac{7}{6}bc + \dfrac{2}{7} =$

(11) $\dfrac{4}{3}x - 2a + \left(-\dfrac{1}{6}a\right) - \dfrac{7}{3}x =$

(5) $-\dfrac{1}{2}ab - \dfrac{5}{6} - \dfrac{3}{4}ab =$

(12) $\dfrac{7}{2}gh - \dfrac{1}{3}fg + \left(-\dfrac{3}{2}gh\right) - \left(-\dfrac{8}{3}fg\right)$

$=$

(6) $-5x - 3 + 8 - 2x =$

(13) $-\dfrac{9}{8}c - \dfrac{3}{5}h + \dfrac{7}{2}c - \left(-\dfrac{3}{10}h\right)$

$=$

(7) $-2x + 5x - 3 - 8 =$

> In your answer, write the variables in alphabetical order.

2 **Simplify each expression.**

(1) $3x^2 + 4x^2 - 8x = \boxed{}\,x^2 - 8x$

 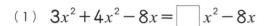

You can combine like terms.

(2) $5x^2 - 2x^2 - 10x + 3x =$

(3) $-5 + 3y^2 + \dfrac{7}{2} - \dfrac{3}{4}y^2 =$

(4) $a^2 + bc - \dfrac{1}{2}bc - \dfrac{10}{3}a^2 =$

(5) $2s + (-v^3) - 3s - (-7v^3) =$

(6) $-\dfrac{7}{3}w + 4n^2 - (-n^2) + (-w)$

$=$

(7) $2a + 3b + 4c + 6a + 7b + 9c$

$= \boxed{}\,a + \boxed{}\,b + \boxed{}\,c$

(8) $x - 5y - 3x + 2 - \dfrac{1}{2} + \dfrac{3}{4}y$

$=$

(9) $-2c^2 + b - \dfrac{2}{3} + 9b - 4c^2 + \dfrac{3}{2}$

$=$

(10) $2x^2 + 5x + 7x^2 + 6 - 5x - 2 = \boxed{}\,x^2 + \boxed{}$

(11) $-(-8y^2) + 10 - 5y^2 + \left(-\dfrac{5}{2}\right) + \dfrac{9}{2}$

$=$

(12) $-5abc - (-2be) + 6be + \dfrac{9}{4}abc - \dfrac{5}{2}be$

$= \boxed{}\,abc + \boxed{}\,be$

You make this look simple!

23

Simplifying Algebraic Expressions

Date   /   /

Name

Score   /100

■ The Answer Key is on page 90.

1 Simplify each expression.

5 points per question

Examples

$$(x-2y)+(3x+4y)=x-2y+3x+4y=4x+2y$$

$$(6x-9y)+(-2x-8y)=6x-9y-2x-8y=4x-17y$$

(1) $(3x+4y)+(x-6y)$

  =

(2) $(a-b)+(-3a-7b)$

  =

(3) $(7g-h)+(-6g+4h)$

  =

(4) $(-m+3n)+(-m+3n)$

  =

(5) $\left(\dfrac{1}{2}a+2b\right)+\left(-2a+\dfrac{3}{2}b\right)$

  =

(6) $\left(-\dfrac{2}{3}k-\dfrac{1}{4}l\right)+\left(-\dfrac{1}{2}k+l\right)$

  =

(7) $\left(8x-\dfrac{3}{2}\right)+(-9x+4)$

  =

(8) $\left(-\dfrac{1}{3}s-\dfrac{1}{4}t\right)+\left(-\dfrac{1}{2}s-\dfrac{3}{5}t\right)$

  =

© Kumon Publishing Co., Ltd.

2 **Simplify each expression.**

10 points per question

(1) $(3a-2b+6c)+(-a-3b+4c)$

=

(2) $(-2x-y+3z)+(5x-4y-2z)$

=

(3) $\left(a+2c-\dfrac{2}{3}d\right)+\left(a-\dfrac{3}{4}c+\dfrac{2}{3}d\right)$

=

(4) $\left(-\dfrac{3}{4}b+\dfrac{5}{2}c^2+\dfrac{1}{3}de\right)+\left(-\dfrac{1}{4}c^2-\dfrac{5}{4}de-2b\right)$

=

> It's okay to leave your answers as improper fractions when there is a variable.

(5) $(2a+9b+3c-5)+(3a-4b-6c+7)$

=

(6) $\left(-f+\dfrac{9}{4}g-\dfrac{5}{3}h+e\right)+\left(\dfrac{3}{2}h-\dfrac{1}{2}f-\dfrac{3}{4}e-\dfrac{3}{4}g\right)$

=

Try to concentrate. Some questions are long.

Simplifying Algebraic Expressions

Level ☆

Date / /

Name

Score /100

■ The Answer Key is on page 90.

1 Write either + or − in each of the boxes.

5 points per question

Example $x-(y+z)=x\ \boxed{-}\ y\ \boxed{-}\ z$

(1) $x-(y-z)=x\ \boxed{}\ y\ \boxed{}\ z$

(5) $x+(-y-z)=x\ \boxed{}\ y\ \boxed{}\ z$

(2) $a+(-b+c)=a\ \boxed{}\ b\ \boxed{}\ c$

(6) $a-(-b-c)=a\ \boxed{}\ b\ \boxed{}\ c$

(3) $-g-(-h-i)=-g\ \boxed{}\ h\ \boxed{}\ i$

(7) $g-(h+2i)=g\ \boxed{}\ h\ \boxed{}\ 2i$

(4) $-l-(-m+n)=-l\ \boxed{}\ m\ \boxed{}\ n$

(8) $l+(-3m-n)=l\ \boxed{}\ 3m\ \boxed{}\ n$

2   **Simplify each expression.**

6 points per question

Example	$5x-(2x+4)=5x-2x-4=3x-4$

Write the correct sign when removing the parentheses.

(1) $9x-(4x-3)=$

(2) $9x-(-4x-3)=$

(3) $9x-(-4x+3)=$

(4) $-5x-(3x-7)=$

(5) $-5x-(-3x+7)=$

(6) $-5x+(-3x-7)=$

(7) $4a+\left(\dfrac{5}{2}-\dfrac{1}{2}a\right)=$

(8) $-4a-\left(\dfrac{5}{2}-\dfrac{1}{2}a\right)=$

(9) $-\dfrac{3}{2}b-(-a+b)=$

(10) $-\dfrac{7}{3}y-\left(-y-\dfrac{3}{4}x\right)=$

Fantastic!

Simplifying Algebraic Expressions

 Level ☆

Score

/ 100

Date / /

Name

■ The Answer Key is on page 90.

1 **Simplify each expression.**

5 points per question

| Example | $(9x+4y)-(2x-6y)=9x+4y-2x+6y=7x+10y$ |

Remember to write a or $-a$, not $1a$ or $-1a$.

(1) $(5x-4y)-(3x-6y)$

=

(6) $(a-2b)-(3a-b)$

=

(2) $(-5x+4y)-(-3x+6y)$

=

(7) $(a+2b)-(3a+b)$

=

(3) $(5x+4y)-(-3x+6y)$

=

(8) $(-a+2b)-(3a-b)$

=

(4) $(-5x-4y)-(-3x-6y)$

=

(9) $(-a-2b)-(-3a+b)$

=

(5) $(-5x-4y)-(3x+6y)$

=

(10) $(-a+2b)-(3a+b)$

=

© Kumon Publishing Co., Ltd.

2 **Simplify each expression.**

5 points per question

(1) $(2a-3b)+(4a-9b)$

=

(2) $(2a-3b)-(4a-9b)$

=

(3) $(-2a+3b)-(-4a-9b)$

=

(4) $(-2a-3b)+(-4a+9b)$

=

(5) $(-3x+2y)+(2x+2y)$

=

(6) $\left(-\dfrac{1}{2}f+3g\right)-\left(\dfrac{1}{3}g+\dfrac{1}{4}f\right)$

=

(7) $\left(-\dfrac{1}{3}y-\dfrac{2}{5}z\right)-\left(\dfrac{3}{5}y-\dfrac{1}{4}z\right)$

=

(8) $\left(\dfrac{3}{4}m-\dfrac{7}{3}n\right)-\left(-\dfrac{1}{3}m-\dfrac{3}{4}n\right)$

=

(9) $\left(-\dfrac{5}{2}a+\dfrac{3}{2}b\right)-\left(-\dfrac{7}{5}b+\dfrac{5}{4}a\right)$

=

(10) $\left(-\dfrac{6}{5}x-\dfrac{3}{8}y\right)-\left(-\dfrac{5}{2}y-\dfrac{8}{3}x\right)$

=

Terrific thinking!

Simplifying Algebraic Expressions

Level ☆

Score

/100

Date   /   /

Name

■ The Answer Key is on page 90.

1 **Simplify each expression.**

5 points per question

Examples

$$
\begin{array}{r}
2x + 5y \\
+)\,4x + 7y \\
\hline
6x + 12y
\end{array}
$$

The above example is the same
as $(2x+5y)+(4x+7y)$

$$
\begin{array}{r}
6a + 11b \\
-)\,2a + 8b \\
\hline
4a + 3b
\end{array}
$$

The above example is the same
as $(6a+11b)-(2a+8b)$

(1)
$$
\begin{array}{r}
3x + 2y \\
+)\ \ x + 4y \\
\hline
4x + 6y
\end{array}
$$

 This is another way to combine like terms.

(5)
$$
\begin{array}{r}
3b + 7c \\
-)\ \ b - 6c \\
\hline
\end{array}
$$

(2)
$$
\begin{array}{r}
5x + y \\
+)\ \ x - 5y \\
\hline
\end{array}
$$

(6)
$$
\begin{array}{r}
x - 9y \\
-)\,-5x + 6y \\
\hline
\end{array}
$$

(3)
$$
\begin{array}{r}
2x - 3y \\
+)\,4x - 2y \\
\hline
\end{array}
$$

(7)
$$
\begin{array}{r}
-\frac{1}{3}s + 3t \\
-)\,-\frac{1}{3}s - 2t \\
\hline
\boxed{}\,t
\end{array}
$$

(4)
$$
\begin{array}{r}
-\frac{5}{7}x + 2y \\
+)\ \ \frac{1}{7}x - 5y \\
\hline
\end{array}
$$

(8)
$$
\begin{array}{r}
\frac{7}{4}a - 6b \\
-)\,-\frac{1}{4}a - 6b \\
\hline
\boxed{}\,a
\end{array}
$$

2 Simplify each expression.

6 points per question

(1)
$$
\begin{array}{r}
3a \qquad + c \\
+)\ 7a + 8b + 4c \\
\hline
10a + 8b + 5c
\end{array}
$$

(6)
$$
\begin{array}{r}
17x + 5y \\
-)\ \ 4x + 2y - 8z \\
\hline
13x + 3y + \square\, z
\end{array}
$$

Pay close attention to the signs.

(2)
$$
\begin{array}{r}
9a - 3b \\
+)\ 2a + 7b - 5c \\
\hline
\end{array}
$$

(7)
$$
\begin{array}{r}
5x \qquad - 9z \\
-)\ 3x - 4y \\
\hline
\end{array}
$$

(3)
$$
\begin{array}{r}
-4x \qquad + z \\
+)\ -3x - 2y \\
\hline
\end{array}
$$

(8)
$$
\begin{array}{r}
-2b + 3c \\
-)\ -a + \ \ b - 5c \\
\hline
\end{array}
$$

(4)
$$
\begin{array}{r}
-\dfrac{3}{2}x^2 - \dfrac{1}{3}x + \dfrac{1}{4} \\
+)\ \qquad \dfrac{1}{5}x - 3 \\
\hline
\end{array}
$$

(9)
$$
\begin{array}{r}
\dfrac{1}{2}x^2 - \dfrac{1}{3}xy + \dfrac{1}{4}y^2 \\
-)\ \dfrac{1}{3}x^2 \qquad - \dfrac{1}{5}y^2 \\
\hline
\end{array}
$$

(5)
$$
\begin{array}{r}
-9x^2 + \dfrac{3}{2}xy - \dfrac{1}{3}y^2 \\
+)\ 9x^2 - \dfrac{3}{2}xy + \dfrac{1}{3}y^2 \\
\hline
\square
\end{array}
$$

(10)
$$
\begin{array}{r}
\dfrac{7}{12}a^2 - 2a \\
-)\ \qquad -\dfrac{2}{3}a - 5 \\
\hline
\end{array}
$$

You're doing a nice job.

31

Simplifying Algebraic Expressions

Level ☆☆

Score

/100

■ The Answer Key is on page 90.

1 Simplify each expression by using the distributive property.   4 points per question

Distributive Property	$a \times b$ $a(b+c)=ab+ac$ $a \times c$	The **distributive property** allows you to simplify the expression by multiplying the terms and removing the parentheses.

(1) $c(d+e)=cd+\boxed{}$

(2) $x(y+z)=\boxed{}+\boxed{}$

(3) $a(b-c)=\boxed{}-\boxed{}$

Be careful to use the correct sign!

(4) $2(x-y)=2x-2y$

(5) $-3(-s+t)=$

(6) $3x+2(x+y)=3x+\boxed{}x+2y=$

(7) $4x-3(x-y)=$

(8) $-2x-(4x-6y)=$

(9) $-\dfrac{1}{2}a+2\left(-a+\dfrac{1}{3}b\right)=$

(10) $\dfrac{3}{4}m-\left(-\dfrac{1}{3}n-\dfrac{1}{2}m\right)=$

2 **Simplify each expression.**

Example $2(3x+4y)+5(x-6y)=6x+8y+5x-30y=11x-22y$

(1) $3(2x-4y)+2(3x-5y)$

$=6x-12y+\boxed{}x-\boxed{}y=$

(2) $4(4x-y)+6(-2x+y)$

$=$

(3) $-3(-3x+5y)+(-x+4y)$

$=$

(4) $6\left(\dfrac{1}{2}a-\dfrac{1}{4}b\right)+2\left(-\dfrac{1}{3}a+\dfrac{1}{4}b\right)$

$=$

(5) $\dfrac{1}{2}(-3a^2-4a)+\dfrac{1}{3}(3a-5a^2)$

$=$

(6) $\dfrac{5}{4}(2x^2+4)+\dfrac{5}{2}(3x^2+8)$

$=$

(7) $4(5x+6y)-2(3x+8y)$

$=$

(8) $2(-3x-3y)-(-5x+7y)$

$=$

(9) $-(-g-h)-4(2g-3h)$

$=$

(10) $3\left(-\dfrac{1}{2}a+b\right)-6\left(a-\dfrac{3}{4}b\right)$

$=$

(11) $\dfrac{1}{2}\left(-5c+\dfrac{1}{3}d\right)-\dfrac{1}{3}\left(\dfrac{1}{4}d-\dfrac{3}{5}c\right)$

$=$

(12) $2(2x+3y-z)-\dfrac{1}{2}\left(-x-y+\dfrac{1}{3}z\right)$

$=$

Simply great work!

Simplifying Algebraic Expressions

Level ☆☆

Date / /

Name

Score

/100

■ The Answer Key is on page 91.

1 **Simplify each expression.**

5 points per question

Examples

$$\frac{2x+3}{7}+\frac{4x+10}{7}=\frac{(2x+3)+(4x+10)}{7}=\frac{2x+3+4x+10}{7}=\frac{6x+13}{7}$$

$$\frac{8x+11y}{9}-\frac{4x-2y}{9}=\frac{(8x+11y)-(4x-2y)}{9}=\frac{8x+11y-4x+2y}{9}=\frac{4x+13y}{9}$$

(1) $\quad\dfrac{6x+9}{4}+\dfrac{x-6}{4}=\dfrac{(\boxed{})+(\boxed{})}{4}$

$=$

(2) $\quad\dfrac{3x-2}{5}+\dfrac{-6x+9}{5}=$

(3) $\quad\dfrac{5x+11}{3}-\dfrac{3x+4}{3}$

$=\dfrac{(\boxed{})-(\boxed{})}{3}=$

(4) $\quad\dfrac{-3x+5}{6}-\dfrac{-4x-9}{6}=$

(5) $\quad-\dfrac{4x+7}{2}-\dfrac{3x-2}{2}$

$=\dfrac{-(\boxed{})-(\boxed{})}{2}=$

(6) $\quad-\dfrac{11x-8}{7}+\dfrac{-3x+4}{7}=$

(7) $\quad\dfrac{3x+5}{4}-\dfrac{9x}{4}=$

(8) $\quad-\dfrac{-x+3}{8}-\dfrac{5}{8}=$

② Simplify each expression.

Example
$$\frac{x-4}{3}+\frac{x-1}{6}=\frac{2(x-4)+(x-1)}{6}=\frac{3x-9}{6}=\frac{x-3}{2}$$

In this example, the last step is to reduce by using the common factor 3.

(1) $\dfrac{2x+3}{3}+\dfrac{4x+5}{4}$

$$=\frac{\boxed{}(2x+3)+\boxed{}(4x+5)}{12}$$

$=$

(2) $\dfrac{4x-5}{2}+\dfrac{3x+2}{3}=$

(3) $\dfrac{-6x+2}{5}+\dfrac{4x-3}{2}=$

(4) $\dfrac{x+3}{4}+\dfrac{2x+5}{6}=$

$$=\frac{\boxed{}(x+3)+\boxed{}(2x+5)}{12}=$$

(5) $\dfrac{2x-5}{6}-\dfrac{-3x-4}{9}=$

(6) $\dfrac{x-3}{2}+\dfrac{5x+7}{4}=$

(7) $\dfrac{-7x+4}{6}-\dfrac{x+1}{2}=$

(8) $-\dfrac{2x-1}{3}+\dfrac{7x-8}{9}=$

(9) $\dfrac{x+5}{2}+\dfrac{7x-11}{6}=$

Reduce your answer!

(10) $\dfrac{2x-9}{15}-\dfrac{x-3}{3}=$

 You can use the least common multiple (LCM) of the denominators.

Keep doing your best!

Solving Equations

Date   /   /

Name

Level ☆☆

Score

/100

■ The Answer Key is on page 91.

1   **Solve each equation.**   2 points per question

Examples

$$x - 3 = 5$$
$$x = 5 + 3$$
$$x = 8$$

$$2 + x = 6$$
$$x = 6 - 2$$
$$x = 4$$

- To solve equations, think "$x =$ a number," and find that number. For questions such as the examples above, add or subtract from both sides of the equation to find the value of x. To find the value of x in the first example, we add 3 to both sides of the equation: $x - 3 \underline{+ 3} = 5 \underline{+ 3}$.
- You can check if your answer is correct by substituting your value for x into the original equation. For example, if you substitute $x = 8$ into the original equation, the value is $8 - 3 = 5$. Therefore, we know that the answer $x = 8$ is correct.

(1) $x - 2 = 7$

$$x = 7 + \boxed{}$$

$$x =$$

(2) $-5 + x = -8$

$$x = -8 + \boxed{}$$

$$x = \boxed{}$$

(3) $4 + x = 1$

(4) $-10 + x = -2$

(5) $x + 9 = 2$

(6) $x + 5 = -1$

(7) $x - 6 = 9$

(8) $3 + x = 6$

(9) $-10 + x = -4$

(10) $-1 + x = -1$

2 **Solve each equation.**

(1) $x + 3 = 7$

(2) $x + 6 = 2$

(3) $5 + x = 0$

(4) $-3 + x = -3$

(5) $x - 0.6 = 2$

(6) $1 + x = 2.5$

(7) $x - 2.9 = 1.2$

(8) $x + 1.9 = -1.6$

(9) $-\dfrac{2}{3} + x = -2$

(10) $x - 3 = -\dfrac{3}{4}$

(11) $5 + x = -\dfrac{3}{2}$

(12) $x - \dfrac{1}{2} = \dfrac{1}{3}$

(13) $\dfrac{1}{4} + x = \dfrac{1}{5}$

(14) $x - \dfrac{3}{4} = \dfrac{5}{2}$

(15) $-\dfrac{9}{4} + x = -\dfrac{3}{2}$

(16) $x + \dfrac{5}{6} = \dfrac{4}{3}$

Wow! You solved them!

Solving Equations

Date / /   Name

Level

Score

/100

■ The Answer Key is on page 91.

1 **Solve each equation.**

5 points per question

Examples

$$2x = 8$$
$$x = 8 \times \frac{1}{2}$$
$$x = 4$$

$$\frac{1}{3}x = -5$$
$$x = -5 \times 3$$
$$x = -15$$

For questions such as the examples above, multiply both sides of the equation to find the value of x. For example, to find the value of x in the first equation, multiply both sides by the **reciprocal** of 2:

$$\frac{1}{2} \times 2x = 8 \times \frac{1}{2}$$

A reciprocal is a fraction flipped upside down.

(1) $5x = 30$

$$x = 30 \times \frac{1}{\boxed{}}$$

$$x =$$

(2) $2x = 14$

(3) $-x = 6$

$$x = 6 \times \left(-\boxed{}\right)$$

$$x =$$

(4) $-3x = 18$

$$x = 18 \times \left(-\boxed{}\right)$$

$$x =$$

(5) $-2x = 7$

(6) $\frac{1}{2}x = 6$

(7) $-\frac{1}{5}x = 2$

(8) $\frac{1}{3}x = \frac{1}{4}$

(9) $-\frac{1}{6}x = 3$

Think of the question as "$x = a$ number."

(10) $\frac{1}{6}x = \frac{3}{4}$

2 Solve each equation.

Example

$$-\frac{2}{3}x = 8$$

$$x = 8 \times \left(-\frac{3}{2}\right)$$

$$x = -12$$

(1) $-\frac{3}{4}x = 6$

(6) $\frac{2}{3}x = -\frac{1}{8}$

(2) $\frac{3}{5}x = 9$

(7) $-\frac{2}{9}x = \frac{1}{12}$

(3) $-\frac{5}{6}x = -10$

(8) $\frac{2}{3}x = \frac{4}{5}$

(4) $\frac{2}{3}x = 5$

(9) $-\frac{8}{11}x = -\frac{6}{7}$

(5) $-\frac{6}{7}x = 8$

(10) $-\frac{7}{2}x = \frac{5}{4}$

You're getting the hang of this!

Solving Equations

Level ★★

Date / /

Name

Score /100

■ The Answer Key is on page 92.

1 **Solve each equation.**

6 points per question

Examples

$$2x - 5 = 7$$
$$2x = 7 + 5$$
$$2x = 12$$
$$x = 12 \times \frac{1}{2}$$
$$x = 6$$

$$\frac{2}{3}x + 7 = -1$$
$$\frac{2}{3}x = -1 - 7$$
$$\frac{2}{3}x = -8$$
$$x = -8 \times \frac{3}{2}$$
$$x = -12$$

(1) $3x - 2 = 10$

$$3x = 10 + \boxed{}$$
$$3x = \boxed{}$$
$$x = \boxed{} \times \frac{1}{\boxed{}}$$
$$x =$$

(2) $-5x - 8 = 7$

(3) $5 - 8x = -1$

(4) $6x - 4 = 10$

(5) $-7x + 1 = 8$

(6) $\frac{3}{5}x - 4 = 2$

$$\frac{3}{5}x = 2 + \boxed{}$$
$$\frac{3}{5}x = \boxed{}$$
$$x = \boxed{} \times \frac{5}{\boxed{}}$$
$$x =$$

2 Solve each equation.

Examples

$$6x = 4x - 10$$
$$6x - 4x = -10$$
$$2x = -10$$
$$x = -10 \times \frac{1}{2}$$
$$x = -5$$

$$10x + 6 = 8x$$
$$10x - 8x + 6 = 0$$
$$2x + 6 = 0$$
$$2x = -6$$
$$x = -6 \times \frac{1}{2}$$
$$x = -3$$

(1) $5x = 2x + 6$

(5) $6x - 4 = 8x$

(2) $7x = x - 2$

(6) $-3x - 1 = 3x$

(3) $4x = 10x + 5$

(7) $\frac{1}{3}x - 4 = \frac{1}{2}x$

(4) $3x = -2x + 4$

(8) $-\frac{1}{4}x - 3 = -\frac{3}{2}x$

Don't give up!

Solving Equations

Date / /

Name

Level
☆☆

Score
/100

■ The Answer Key is on page 92.

1 **Solve each equation.**

10 points per question

Example

$$5x + 7 = 3x + 13$$

$$5x - 3x + 7 = 13$$

$$2x + 7 = 13$$

$$2x = 13 - 7$$

$$2x = 6$$

$$x = 6 \times \frac{1}{2}$$

$$x = 3$$

In the example:
- "$3x$" on the right of the equal sign has been transposed to "$-3x$" on the left
- "$+7$" on the left of the equal sign has been transposed to "-7" on the right

Transposition (or **Inverse Operations**) is when we move a term to the other side of the equation and change the sign.

(1) $2x - 7 = -6x + 9$

$2x + 6x - \boxed{} = 9$

$\boxed{}x - \boxed{} = 9$

(3) $5x - 1 = 7x + 11$

(2) $x - 5 = 2x - 1$

(4) $-2x - 4 = 2x + 8$

2 Solve each equation.

(1) $3x + 8 = -2x + 7$

(4) $\dfrac{1}{2}x + 4 = \dfrac{1}{3}x + 1$

(2) $2x - 8 = -11x - 10$

(5) $-\dfrac{2}{5}x + \dfrac{1}{3} = -\dfrac{1}{2}x + \dfrac{1}{4}$

(3) $6x - 6 = 3x + 1$

(6) $\dfrac{5}{3}x - 2 = \dfrac{1}{2}x - \dfrac{2}{3}$

You got to the answer step-by-step!

Date   /   /

Name

■ The Answer Key is on page 92.

1 **Solve each equation.**

6 points per question

Example $7x - 6 = 5x + 10$

$7x - 5x = 10 + 6$

$2x = 16$

$x = 8$

(1) $9x - 8 = 6x + 4$

$9x \boxed{} 6x = 4 \boxed{} 8$

(4) $-2x + 11 = 2x + 3$

(2) $7x + 2 = 2x - 13$

$7x \boxed{} 2x = -13 \boxed{} 2$

(5) $-3x - 2 = 5x + 6$

(3) $7x - 2 = -5x - 14$

(6) $4x + 2 = -2x - 10$

2 **Solve each equation.**

8 points per question

(1) $-9x+2=-6x-7$

(2) $8x-6=9x-2$

(3) $5x+3=3x-6$

(4) $5x-7=-3x+9$

(5) $6x+2=4x+3$

(6) $-2x-5=x-7$

(7) $2x-2=\dfrac{1}{2}x+6$

(8) $x+\dfrac{2}{3}=-\dfrac{1}{3}x-\dfrac{1}{6}$

Your work is getting better and better!

© Kumon Publishing Co., Ltd. 45

Solving Equations

23

Date / /

Name

Level

Score

/100

■ The Answer Key is on page 92.

1 **Solve each equation, and check your answer.**   10 points per question

Example	$5x - 1 = 3x - 7$	Check your answer!

$$5x - 1 = 3x - 7$$
$$5x - 3x = -7 + 1$$
$$2x = -6$$
$$x = -3$$

Check your answer!

Left side $= 5 \times (-3) - 1 = -15 - 1 = -16$

Right side $= 3 \times (-3) - 7 = -9 - 7 = -16$

If the value on the left side of the equal sign matches the value on the right, then your answer is correct!

(1)　$2x - 6 = -3x + 4$

$2x + \boxed{} = 4 + \boxed{}$

Check!

Left side $= 2 \times \boxed{} - 6 =$

Right side $= -3 \times \boxed{} + 4 =$

(2)　$4x + 7 = 9 + 5x$

Check!

Left side $=$

Right side $=$

2 **Solve each equation, and check your answer.**

20 points per question

(1) $3x - 3 = -3x - 15$

Check!

Left side =

Right side =

(2) $1 + \dfrac{2}{3}x = x - 1$

Check!

Left side =

Right side =

(3) $\dfrac{3}{4}x - 1 = -\dfrac{3}{2}x + 2$

Check!

Left side =

Right side =

(4) $-\dfrac{1}{2}x + \dfrac{1}{5} = -3x + \dfrac{1}{2}$

Check!

Left side =

Right side =

By checking, you're sure to
get every answer correct!

Solving Equations

Date / /

Name

■ The Answer Key is on page 92.

1 **Solve each equation.**

6 points per question

Example	$3(4x+6)=2(5x+7)$	To simplify, remove the parentheses.
	$12x+18=10x+14$	
	$12x-10x=14-18$	
	$2x=-4$	
	$x=-2$	

(1) $2(x-5)=-4$

$2x-\boxed{}=-4$

(4) $-2(x+1)=-6(x-1)$

(2) $3x=4(x+2)$

(5) $-(x-5)=3(7-x)$

(3) $-(2-4x)=-(x-3)$

(6) $4(-6+x)=-3(2x-7)$

2 Solve each equation.

8 points per question

(1) $4+2(x+4)=-x$

$4+\boxed{}x+\boxed{}=-x$

(5) $-(x+5)=2(2x-3)-7$

(2) $2-3(2x+3)=x$

(6) $6\left(\dfrac{1}{2}x+1\right)-7=-3\left(-\dfrac{1}{2}x-1\right)$

(3) $-x-(3x-2)=14$

(7) $2\left(-\dfrac{3}{4}x-\dfrac{1}{2}\right)-2x=-\left(-\dfrac{1}{3}+x\right)$

(4) $4(-x-2)-3x=-x$

(8) $-\dfrac{2}{3}(3+2x)+4=-(1+x)-\dfrac{1}{2}x$

You're a champ!

Date   /   /

Name

■ The Answer Key is on page 92.

1 **Solve each equation.**

10 points per question

Example

$$\frac{1}{3}x + \frac{1}{6} = \frac{1}{2}x + \frac{5}{6}$$

$$\left(\frac{1}{3}x + \frac{1}{6}\right) \times 6 = \left(\frac{1}{2}x + \frac{5}{6}\right) \times 6$$

$$2x + 1 = 3x + 5$$

$$2x - 3x = 5 - 1$$

$$-x = 4$$

$$x = -4$$

To remove the denominator and solve for x, multiply each side by the LCM of the denominators—in this example, it is 6.

(1) $\frac{1}{4}x + \frac{1}{2} = \frac{1}{6}x + \frac{1}{3}$

Multiply each side by $\boxed{}$:

$$\left(\frac{1}{4}x + \frac{1}{2}\right) \times \boxed{} = \left(\frac{1}{6}x + \frac{1}{3}\right) \times \boxed{}$$

(3) $5 - \frac{1}{4}x = 7 - \frac{1}{6}x$

(2) $\frac{7}{10}x + \frac{2}{5} = \frac{7}{20} + \frac{3}{4}x$

(4) $-\frac{5}{6}x - 2 = \frac{1}{2} - \frac{3}{8}x$

② Solve each equation.

Example

$$\frac{x+1}{3} = \frac{2x-7}{5}$$

$$\frac{x+1}{3} \times 15 = \frac{2x-7}{5} \times 15$$

$$5(x+1) = 3(2x-7)$$

$$5x+5 = 6x-21$$

$$-x = -26$$

$$x = 26$$

(1) $\dfrac{x-2}{4} = \dfrac{3x-7}{10}$

$$\frac{x-2}{4} \times \boxed{} = \frac{3x-7}{10} \times \boxed{}$$

(4) $\dfrac{x}{3} - \dfrac{3-2x}{6} = \dfrac{5}{2}x$

You can always check your answer by plugging the value of x into the original equation.

(2) $\dfrac{-6x-2}{5} = \dfrac{4-3x}{2}$

(5) $-\dfrac{6x-7}{5} + 3 = \dfrac{4-3x}{2}$

(3) $\dfrac{-2x-3}{4} = \dfrac{-6-5x}{6}$

(6) $\dfrac{-3-5x}{4} - 2 = \dfrac{2x-1}{6} - 1$

Extraordinary job!

26 Word Problems with Equations

Level ☆☆

Date / /

Name

Score
/100

■ The Answer Key is on page 93.

1 Write each sentence as an equation, and then solve.

6 points per question

(1) x added to 3 is 7.

$$3 + x = \boxed{}$$

(6) x equals 6 times -2.

$$x = 6 \times (\boxed{})$$

(2) Adding $\frac{1}{2}$ and x is -3.

(7) $\frac{3}{4}$ times x equals 6.

(3) Subtracting 2 from x equals 10.

$$\boxed{} - 2 = 10$$

(8) x divided by 5 equals 4.

$$\frac{x}{\boxed{}} = 4$$

(4) Subtracting $\frac{1}{2}$ from x equals $\frac{1}{3}$.

(9) x divided by 3 equals $\frac{2}{5}$.

(5) Subtracting x from 5 equal 4.

$$5 - x = \boxed{}$$

(10) x divided by 12 is $\frac{2}{3}$.

2 **Write each word problem as an equation, and then solve.**

(1) Patty has a bag of yo-yos that weighs 76 ounces. If each yo-yo weighs 4 ounces, how many yo-yos are in Patty's bag?

If there are x yo-yos,

$4 \times x = \boxed{}$

⟨**Ans.**⟩ _____ yo-yos

(2) Nathalie purchases a crate of potatoes that weighs 192 ounces. If there are 24 potatoes in the crate, how much does each potato weigh?

If each potato weighs x ounces,

⟨**Ans.**⟩ _____

(3) Dr. Mary orders a box of apples and oranges that weighs 39 ounces. Each apple weighs 5 ounces and each orange weighs 4 ounces. If there are 3 apples in the box, how many oranges are there?

If there are x oranges,

$5 \times 3 + \boxed{} \times x = 39$

Don't forget to write the units!

⟨**Ans.**⟩ _____

(4) Eriko buys a bag of peanuts and walnuts that weighs 315 grams. There are 25 peanuts and 40 walnuts in the bag. If each peanut weighs 3 grams, how much does each walnut weigh?

If each walnut weighs x grams,

⟨**Ans.**⟩ _____

You're a whiz with word problems!

Date / /

Name

Score /100

■ The Answer Key is on page 93.

1 Write each word problem as an equation.

10 points per question

(1) There are a total of 12 pencils and pens in Hideyo's backpack. If there are x pencils, how many pens are there?

⟨Ans.⟩ $\left(\boxed{} - x \right)$ pens

(2) There are a total of 30 crayons and pens on Trent's desk. If there are x pens, how many crayons are there?

⟨Ans.⟩

(3) In Christy's pet shop, there are 2 more dogs than cats. If there are x dogs, how many cats are there?

⟨Ans.⟩

(4) There are 25 students in Evan's class. If there are x boys, how many girls are there?

⟨Ans.⟩

(5) In Kevin's class, there are 5 more girls than boys. If there are x boys, how many girls are there?

⟨Ans.⟩

(6) In Maurice's class, there are also 5 more girls than boys. If there are x girls, how many boys are there?

⟨Ans.⟩

2 **Write the word problem as two possible equations, and then solve.** 20 points for completion

There are 24 students in Steve's swimming class. There are 6 more girls than boys in the pool. How many boys are in the pool? How many girls are in the pool?

Method #1:

If there are x boys in the pool,

then there are $(x + \boxed{})$ girls in the pool.

$x + (x + \boxed{}) = 24$

⟨**Ans.**⟩ _____ boys

_____ girls

Method #2:

If there are x girls in the pool,

then there are $(x - \boxed{})$ boys in the pool.

$x + (\boxed{}) = 24$

⟨**Ans.**⟩

_____

3 **Write the word problem as two possible equations, and then solve.** 20 points for completion

Debby bought 31 apples and bananas. There are 9 more apples than bananas. How many apples did she buy? How many bananas did she buy?

⟨**Ans.**⟩

_____

⟨**Ans.**⟩

_____

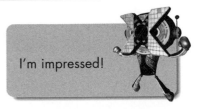

I'm impressed!

■The Answer Key is on page 93.

1 **Write each word problem as an equation, and then solve.**

20 points per question

(1) Jeanine has 22 markers, and Mary has 8 markers. Jeanine gives Mary x markers, so they will each have the same number of markers. How many markers did Jeanine give Mary?

Jeanine gives Mary x markers,

therefore Jeanine has a new amount of $(22 - \boxed{})$ markers,

and Mary has a new amount of $(\boxed{} + x)$ markers.

Because they have the same number of markers,

$22 - \boxed{} = \boxed{} + x$

⟨Ans.⟩ _____

(2) John has 20 coins, and Laura has 6 coins. John gives Laura x coins, so John will only have 4 more coins than Laura. How many coins did John give Laura?

John gives Laura x coins,

therefore John has a new amount of $(20 - \boxed{})$ coins,

and Laura has a new amount of $(6 + x)$ coins.

Because John has 4 more coins than Laura,

$20 - \boxed{} = (6 + x) + \boxed{}$

⟨Ans.⟩ _____

2 Write each word problem as an equation, and then solve.

20 points per question

(1) Masa is 12 years old. In how many years will Masa be 4 times his current age?

If x is the number of years that pass,

then Masa's new age will be $\left(12 + \boxed{}\right)$ years old.

Because Masa will be 4 times his current age,

$$\left(12 + \boxed{}\right) = \boxed{} \times 12$$

⟨**Ans.**⟩ _____

(2) In 10 years, Mike will be 3 times his current age. What is Mike's current age?

If x is his current age,

then

⟨**Ans.**⟩ _____

(3) Justin is 10 years old, and Alan is 26 years old. In how many years will Alan be twice as old as Justin?

If x is the number of years that pass,

then Justin's new age will be $\left(10 + \boxed{}\right)$ years old,

and Alan's new age will be $\left(26 + \boxed{}\right)$ years old.

Because Alan will be twice as old as Justin,

$$\boxed{} \times \left(10 + \boxed{}\right) = 26 + \boxed{}$$

⟨**Ans.**⟩ _____

These are tricky! Nice job!

Solving Equations

■ The Answer Key is on page 93.

1 Solve each equation for x.

5 points per question

Examples

$x+3=b$

$x=b-3$

Transpose 3 to the right of the equal sign.

$a=b-x$

$x=b-a$

Transpose $-x$ to the left of the equal sign and a to the right.

(1) $x+6=b$

$x=b-\boxed{}$

(6) $-a=-b-x$

(2) $x+1=c$

(7) $c=-d-x$

(3) $x-8=a$

$x=a+\boxed{}$

(8) $-1=b-x$

$x=b+\boxed{}$

(4) $x-4=d$

(9) $2=-y-x$

(5) $x-11=-g$

(10) $-a=-3-x$

In your answer, write the variables in alphabetical order.

© Kumon Publishing Co., Ltd.

2 **Solve each equation for** x.

5 points per question

Examples $2x = a$ $4x = a + b$

$$x = \frac{a}{2}$$ $$x = \frac{a + b}{4}$$

---

(1) $ax = c$

(6) $bx = -c + d$

(2) $-3x = b$

$$x = \frac{b}{\boxed{}}$$

$$x = -\frac{b}{\boxed{}}$$

(7) $cx + 3 = d$

$$cx = d - \boxed{}$$

(3) $-fx = -g$

(8) $ax - 5 = b$

(4) $s = 4x$

$$4x = \boxed{}$$

$$x =$$

(9) $s - 2t = rx$

$$rx = \boxed{}$$

(5) $7 = ax$

(10) $-y + 2z = wx$

Hats off to you!

Solving Equations

Date / /

Name

■ The Answer Key is on page 93.

1 **Solve each equation for x.**

10 points per question

Example

$$\frac{x}{a} + \frac{b}{c} = d$$

$$ac\left(\frac{x}{a} + \frac{b}{c}\right) = acd$$

$$cx + ab = acd$$

$$cx = acd - ab$$

$$x = \frac{acd - ab}{c}$$

To remove the denominator and solve for x, multiply each side by the LCM of the denominators—in this example, it is ac.

(1) $\dfrac{x}{a} = b - \dfrac{c}{d}$

Multiply each side by the LCM of the

denominators, which is ☐.

(4) $\dfrac{x}{c} = \dfrac{d}{b}$

Multiply each side by the LCM of the

denominators, which is ☐.

(2) $\dfrac{x}{6} = -2 - \dfrac{3}{4}x$

(5) $\dfrac{a}{b}x = \dfrac{c}{d}$

(3) $2x + \dfrac{5}{3} = -\dfrac{1}{6}x$

(6) $\dfrac{3f}{g} = -\dfrac{4x}{5h}$

2 **Solve each equation for x.**

(1) $x - a = b$

(2) $2(x - a) = b$

   $x - a = \dfrac{b}{\boxed{}}$

(3) $y(x + 4) = a$

(4) $a(x + b) = c + d$

(5) $\dfrac{1}{3}(x - a) = b$

(6) $\dfrac{1}{5}(x + 3) = -d$

(7) $\dfrac{1}{a}(3 - 4x) = 5$

(8) $\dfrac{a(b - d)}{x} = -M$

You go the extra mile, and it shows!

61

Simultaneous Linear Equations

Level ☆☆☆

Score

Date   /   /

Name

/100

■ The Answer Key is on page 94.

1 **Solve each equation for both variables.**

15 points per question

Example

$$\begin{cases} 4x+5y=7 & \cdots① \\ x+5y=13 & \cdots② \end{cases}$$

Step 1: Number each equation.

$①-②:\quad 4x+5y=7 \quad\cdots①$

$\underline{-)\ x+5y=13\ \cdots②}$

$\qquad 3x\qquad =-6$

$\qquad\quad x=-2$

Step 2: Remove a variable by subtracting one equation from the other.

Substitute this into ② :

$$-2+5y=13$$
$$5y=15$$
$$y=3$$

Step 3: Substitute the value of the variable into either original equation (usually the simpler one) to solve for the other variable.

⟨Ans.⟩ $(x, y)=(-2,\ 3)$

Step 4: Write your answer.

Two linear equations with the same variables are called **simultaneous linear equations**. You can solve for the variables by subtracting one equation from the other, a method known as the **subtraction method**.

(1) $\begin{cases} x+3y=-13 & \cdots① \\ 7x+3y=-19 & \cdots② \end{cases}$

$①-②: -6x=\boxed{}$

$\qquad\quad x=\boxed{}$

Substitute this into ① :

$\boxed{}+3y=-13$

$\qquad 3y=\boxed{}$

$\qquad\ y=\boxed{}$

⟨Ans.⟩ $(x, y)=(\boxed{},\ \boxed{})$

(2) $\begin{cases} 2x-5y=-17 & \cdots① \\ 2x+3y=23 & \cdots② \end{cases}$

$①-\boxed{②} :$

Substitute this into ② :

⟨Ans.⟩

2 **Solve each equation for both variables.**

14 points per question

(1) $\begin{cases} 2x+3y=4 \\ 7x+3y=-1 \end{cases}$

(4) $\begin{cases} -7x-2y=-38 \\ -7x+8y=12 \end{cases}$

⟨Ans.⟩ _____

⟨Ans.⟩ _____

(2) $\begin{cases} 4x+3y=-15 \\ 4x-y=-27 \end{cases}$

(5) $\begin{cases} -5x-4y=17 \\ 3x-4y=-23 \end{cases}$

⟨Ans.⟩ _____

⟨Ans.⟩ _____

(3) $\begin{cases} 5x-6y=4 \\ 3x-6y=0 \end{cases}$

Don't forget to number the equations!

This is complex, so concentrate!

⟨Ans.⟩ _____

Simultaneous Linear Equations

Level ☆☆☆

Score

/100

Date / /

Name

■ The Answer Key is on page 94.

1 **Solve each equation for both variables.**

14 points per question

Example

$$\begin{cases} 5x+4y=26 \quad \cdots ① \\ 3x-4y=22 \quad \cdots ② \end{cases}$$

Step 1: Number each equation.

$$①+②: \quad 5x+4y=26 \quad \cdots ①$$
$$\underline{+) \; 3x-4y=22 \quad \cdots ②}$$
$$8x \qquad =48$$
$$x=6$$

Step 2: Remove a variable by adding one equation to the other.

Substitute this into ① :

$$30+4y=26$$
$$4y=-4$$
$$y=-1$$

Step 3: Substitute the value of the variable into either original equation (usually the simpler one) to solve for the other variable.

⟨Ans.⟩ $(x, y)=(6, -1)$

Step 4: Write your answer.

You can solve for the variables by adding the two equations together, a method known as the
addition method.

(1) $\begin{cases} 5x-3y=13 \\ -x+3y=-5 \end{cases}$

(2) $\begin{cases} -4x+y=17 \\ 4x-8y=4 \end{cases}$

⟨**Ans.**⟩

⟨**Ans.**⟩

64 © Kumon Publishing Co., Ltd.

2 **Solve each equation by adding or subtracting the equations.** 18 points per question

(1) $\begin{cases} 5x + 2y = -22 & \cdots \text{①} \\ 3x - 2y = -10 & \cdots \text{②} \end{cases}$

① ☐ ② :

(3) $\begin{cases} \dfrac{1}{2}x + 4y = 15 \\ \dfrac{1}{2}x - 5y = -12 \end{cases}$

⟨Ans.⟩ _____

(2) $\begin{cases} 6x + 2y = -5 \\ 6x - y = 7 \end{cases}$

(4) $\begin{cases} 2x - \dfrac{1}{3}y = 1 \\ \dfrac{1}{2}x + \dfrac{1}{3}y = \dfrac{3}{4} \end{cases}$

⟨Ans.⟩ _____

⟨Ans.⟩ _____

⟨Ans.⟩ _____

Variables can be fractions, too.

You can solve in two ways now! Wow!

33 Simultaneous Linear Equations

Level ☆☆☆

Date / /

Name

Score

/100

■ The Answer Key is on page 94.

1 Solve each equation.

20 points per question

Example

$$\begin{cases} -3x + y = 9 & \cdots ① \\ -5x - 2y = 26 & \cdots ② \end{cases}$$

$① \times 2 : \qquad -6x + 2y = 18 \ \cdots ③$

$② + ③ : \qquad -5x - 2y = 26 \ \cdots ②$

$$\underline{+) -6x + 2y = 18 \ \cdots ③}$$

$$-11x \qquad = 44$$

$$x = -4$$

Substitute this into ① :

$$12 + y = 9$$

$$y = -3$$

⟨Ans.⟩ $(x, y) = (-4, -3)$

Step 1: Number each equation.

Step 2: Multiply one equation by a constant and number the new equation. In this example, we multiplied equation ① by 2 and called it equation ③.

Step 3: Remove a variable by using the addition or subtraction method.

Step 4: Substitute the value of the variable into either original equation (usually the simpler one) to solve for the other variable.

Step 5: Write your answer.

(1) $\begin{cases} 2x + y = 11 & \cdots ① \\ 5x - 2y = 5 & \cdots ② \end{cases}$

$① \times 2 : 4x + \boxed{}\, y = \boxed{} \cdots ③$

$② + ③ : \qquad 9x = \boxed{}$

$\qquad\qquad x = \boxed{}$

Substitute this into ① :

$$2 \times \boxed{} + y = 11$$

$$y = \boxed{}$$

⟨Ans.⟩ $(x, y) = (\boxed{}, \boxed{})$

(2) $\begin{cases} 8x - 5y = 21 & \cdots ① \\ 2x + 9y = -5 & \cdots ② \end{cases}$

$② \times 4 :$

⟨Ans.⟩

© *Kumon Publishing Co., Ltd.*

2 **Solve each equation, and check your answer.** 30 points per question

┌─ Don't forget! ──────────────────────────────────────┐

You can always check your answer by substituting your answer into the original equations and checking if both equations are satisfied.

(1) $\begin{cases} 2x + 3y = 12 & \cdots ① \\ 5x - y = 13 & \cdots ② \end{cases}$

Check!

① : $2 \times \boxed{} + 3 \times \boxed{} = \boxed{} + \boxed{} = \boxed{}$

The left side of ① matches the right side!

② : $5 \times \boxed{} - \boxed{} = \boxed{} - \boxed{} = \boxed{}$

The left side of ② matches the right side!

⟨Ans.⟩ _____

Therefore, your answer is correct!

(2) $\begin{cases} 6x + 5y = -9 & \cdots ① \\ -2x - 3y = 7 & \cdots ② \end{cases}$

Check!

⟨Ans.⟩ _____

Double checking your work is smart!

Level

Score /100

Date / /

Name

■ The Answer Key is on page 94.

1 **Solve each equation.**

20 points per question

Example

$$\begin{cases} 5x + 2y = 16 & \cdots① \\ 4x + 3y = 17 & \cdots② \end{cases}$$

Step 1: Number each equation.

$①×3:\quad 15x + 6y = 48 \quad\cdots③$

Step 2: Multiply one equation by a constant and number the new equation.

$②×2:\quad 8x + 6y = 34 \quad\cdots④$

Step 3: Multiply the other equation by a constant and number the new equation.

$$③-④:\quad \begin{array}{r} 15x+6y=48 \quad\cdots③ \\ -)\ \ 8x+6y=34 \quad\cdots④ \\ \hline 7x \qquad = 14 \\ x = 2 \end{array}$$

Step 4: Remove a variable by using the addition or subtraction method.

Substitute this into ② :

$$8 + 3y = 17$$
$$3y = 9$$
$$y = 3$$

Step 5: Substitute the value of the variable into either original equation (usually the simpler one) to solve for the other variable.

⟨Ans.⟩ $(x, y) = (2,\ 3)$

Step 6: Write your answer.

(1) $\begin{cases} 5x + 3y = -7 & \cdots① \\ 6x + 4y = -8 & \cdots② \end{cases}$

$①×4:\quad \begin{cases} 20x + 12y = \boxed{} \quad\cdots③ \\ 18x + 12y = \boxed{} \quad\cdots④ \end{cases}$

$②×\boxed{}:$

$③-④:\qquad 2x = \boxed{}$

$\qquad\qquad x = \boxed{}$

Substitute this into ① :

$$5 × \left(\boxed{}\right) + 3y = -7$$
$$3y = \boxed{}$$
$$y = \boxed{}$$

⟨Ans.⟩ $(x, y) = \left(\boxed{},\ \boxed{}\right)$

(2) $\begin{cases} -2x + 3y = -2 & \cdots① \\ 5x - 7y = 6 & \cdots② \end{cases}$

$①×5:\quad \Big\{$

$②×\boxed{}:$

⟨Ans.⟩

2 Solve each equation, and check your answer.

(1) $\begin{cases} 3x - 5y = 8 & \cdots ① \\ -2x + 7y = 2 & \cdots ② \end{cases}$

Check!

① : $3 \times \boxed{} - 5 \times \boxed{} = \boxed{} - \boxed{} =$
The left side of ① matches the right side.

② : $-2 \times \boxed{} + 7 \times \boxed{} = \boxed{} + \boxed{} =$
The left side of ② matches the right side.

Therefore, your answer is correct!

⟨**Ans.**⟩ _____

(2) $\begin{cases} -8x + 3y = 8 & \cdots ① \\ 9x + 5y = -9 & \cdots ② \end{cases}$

⟨**Ans.**⟩ _____

(3) $\begin{cases} 5x - 6y = -12 & \cdots ① \\ 4x - 9y = -11 & \cdots ② \end{cases}$

⟨**Ans.**⟩ _____

Double checking your work takes time, but it's worthwhile!

35

Date   /   /

Name

■ The Answer Key is on page 94.

1 **Rewrite each equation into the form** $ax + by = c$ **and solve for both variables.**

10 points per question

┌─ **Don't forget!** ─────────────────────────────────────┐
│ In order to solve equations, it may be easier to first rearrange and/or simplify the equations into │
│ the form $ax + by = c$. │
└──┘

(1) $\begin{cases} 5y = -2x - 8 & \cdots ① \\ -4y = 3x - 2 & \cdots ② \end{cases}$

(3) $\begin{cases} 7x = 3y - 2 \\ 5x = -26 - 4y \end{cases}$

① becomes : $2x + 5y = -8 \quad \cdots ③$

② becomes : $\boxed{}x - 4y = -2 \quad \cdots ④$

③ × 3 : $\boxed{}x + 15y = -24 \quad \cdots ⑤$

④ × $\boxed{}$: $-6x - 8y = -4 \quad \cdots ⑥$

⑤ + ⑥ : $\boxed{}y = -28$

$y = \boxed{}$

⟨Ans.⟩ _____

⟨Ans.⟩ _____

(2) $\begin{cases} -2y = -3x + 9 \\ 7y - 1 = 4x \end{cases}$

(4) $\begin{cases} 3y = 6x + 3 \\ -9x - 9 = -5y \end{cases}$

⟨Ans.⟩ _____

⟨Ans.⟩ _____

2 **Solve each equation.**

15 points per question

(1) $\begin{cases} 4x + 2y = x + 8y + 18 & \cdots① \\ 5x - 13y = 6x - 9y & \cdots② \end{cases}$

　① becomes : $3x - \boxed{}y = 18$ $\cdots③$

　② becomes : $-x - \boxed{}y = 0$ $\cdots④$

(3) $\begin{cases} -2(3x + 2) = 4 - y \\ 5(x - 3y) = 2(6x - 5y) - 3 \end{cases}$

〈Ans.〉 _____

〈Ans.〉 _____

(2) $\begin{cases} 3(x + 4y) = x + 7y + 2 & \cdots① \\ -9x - 10y + 28 = -12x - 2y & \cdots② \end{cases}$

　① becomes : $\boxed{}x + 5y = 2$ $\cdots③$

　② becomes : $\boxed{}x - \boxed{}y = -28$ $\cdots④$

(4) $\begin{cases} -(3 - 2x) = 4(x - y) \\ 2(x - 4 + y) = -3(y + 3 - 2x) \end{cases}$

〈Ans.〉 _____

〈Ans.〉 _____

Remove parentheses first.

You are getting good at this!

■ The Answer Key is on page 94.

1 Solve each equation.

15 points per question

(1) $\begin{cases} 4x + \dfrac{2}{3}y = -2 & \cdots ① \\ 5x = 7 - 4y & \cdots ② \end{cases}$

Remove the denominator from ① :

①×3 : $12x + \boxed{}\,y = \boxed{}$ $\cdots ③$

Rearrange the terms in ② :

$5x + \boxed{}\,y = 7$ $\cdots ④$

③×2 :

⟨Ans.⟩ _____

(2) $\begin{cases} \dfrac{5}{6}x + \dfrac{1}{4}y = \dfrac{1}{3} & \cdots ① \\ 7x - 4 = -2y - 2 & \cdots ② \end{cases}$

Remove the denominator from ① :

①×12 :

⟨Ans.⟩ _____

(3) $\begin{cases} \dfrac{2}{5}x - 1 = \dfrac{1}{2}y & \cdots ① \\ \dfrac{1}{4}(x - 2y) = \dfrac{2}{3}(x - y) - \dfrac{7}{4} & \cdots ② \end{cases}$

Remove the denominator from ① :

①×10 : $4x - 10 = \boxed{}\,y$

$\therefore 4x - \boxed{}\,y = 10$ $\cdots ③$

Remove the denominator from ② :

②×12 : $3(x - 2y) = \boxed{}(x - y) - 21$

$\therefore -5x + 2y = \boxed{}$ $\cdots ④$

$\therefore$ means "therefore."

⟨Ans.⟩ _____

(4) $\begin{cases} -\dfrac{1}{3}(4x - 5y) + 1 = \dfrac{1}{2}(3y - 2x) \\ \dfrac{2}{5}(4 - 2x) + \dfrac{3}{5}y = -\dfrac{1}{2}x + 1 \end{cases}$

⟨Ans.⟩ _____

2 **Solve each equation.**

20 points per question

(1) $\begin{cases} \dfrac{x-y}{3} + \dfrac{2x-y}{4} = 2 & \cdots① \\[3mm] \dfrac{x-y}{2} + \dfrac{x+y}{3} = \dfrac{7}{6} & \cdots② \end{cases}$

Remove the denominator from ①

and ② by :

①×12 : $4(x-y) + \boxed{}(2x-y) = 24$

②×$\boxed{}$: $3(x-y) + \boxed{}(x+y) = \boxed{}$

∴ $\begin{cases} \boxed{}x - \boxed{}y = 24 \\[2mm] \boxed{}x - y = \boxed{} \end{cases}$

Use ∴ when completely rearranging an equation.

⟨Ans.⟩ _____

(2) $\begin{cases} \dfrac{2x-3y}{4} - \dfrac{x-y}{2} = \dfrac{x+1}{5} & \cdots① \\[3mm] -\dfrac{x+2y+1}{3} + 1 = \dfrac{-3x-2y-1}{2} & \cdots② \end{cases}$

Remove the denominator from ①

and ② by :

⟨Ans.⟩ _____

You went above and beyond!

© Kumon Publishing Co., Ltd.

37

Simultaneous Linear Equations

 Level ☆☆☆

Date / /

Name

Score /100

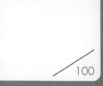

■ The Answer Key is on page 95.

1 **Solve each equation.**

14 points per question

Example

$$\begin{cases} y = 2x - 1 & \cdots ① \\ 4x - 3y = -7 & \cdots ② \end{cases}$$

Step 1: Number each equation.

Substitute ① into ② :

$$4x - 3(2x - 1) = -7$$
$$4x - 6x + 3 = -7$$
$$-2x = -10$$
$$x = 5$$

Step 2: Substitute one equation into the other to remove a variable.

Substitute this into ① :

$$y = 10 - 1$$
$$= 9$$

Step 3: Substitute the value of the variable into either original equation (usually the simpler one) to solve for the other variable.

⟨Ans.⟩ $(x, y) = (5, 9)$

Step 4: Write your answer.

The **substitution method** allows you to substitute one equation into the other equation in order to eliminate one variable and solve for the remaining variable.

(1) $\begin{cases} y = 3x - 2 \\ 2x - y = -1 \end{cases}$

(2) $\begin{cases} x = y + 4 \\ 2x + 3y = -2 \end{cases}$

⟨Ans.⟩ _____

⟨Ans.⟩ _____

2 Solve each equation by using the substitution method.

18 points per question

(1) $\begin{cases} 4x + 5y = 10 & \cdots① \\ x + 3y = -1 & \cdots② \end{cases}$

Rewrite ② : $x = \boxed{}\, y - 1 \quad \cdots③$

Substitute ③ into ① :

$$4\left(\boxed{}\, y - 1\right) + 5y = 10$$

$$\boxed{}\, y = 14$$

$$y = \boxed{}$$

Substitute this into ③ :

$$x = \boxed{} - 1$$

$$x =$$

⟨Ans.⟩ $(x, y) = \left(\boxed{}, \boxed{}\right)$

(2) $\begin{cases} 3x - 5y = -12 \\ x = y \end{cases}$

⟨Ans.⟩

(3) $\begin{cases} 5x + 3y = 8 & \cdots① \\ -4x + 7y = 3 & \cdots② \end{cases}$

Rewrite ① : $y = \dfrac{\boxed{}}{3} \quad \cdots③$

Substitute ③ into ② :

⟨Ans.⟩

(4) $\begin{cases} 7x + 2y = -6 \\ 8x + 5y = 4 \end{cases}$

⟨Ans.⟩

You can always go back and check your answers by plugging the value of each variable into both equations.

There is no substitute for hard work and practice!

© Kumon Publishing Co., Ltd.   75

38

Simultaneous Linear Equations

Date / / .

Name

■ The Answer Key is on page 95.

1 **Solve each equation by using the substitution method.**

10 points per question

(1) $\begin{cases} 2x = -3y + 7 & \cdots ① \\ 3x - 4y = 2 & \cdots ② \end{cases}$

Rewrite ① : $x = \dfrac{-3y + 7}{\boxed{}}$ $\cdots ③$

Substitute ③ into ② :

$3\left(\dfrac{-3y + 7}{\boxed{}}\right) - 4y = 2$

$3(-3y + 7) - 8y = \boxed{}$

$-17y = \boxed{}$

$y = \boxed{}$

Substitute this into ③ :

$x = \dfrac{\boxed{} + 7}{\boxed{}}$

$x =$

⟨Ans.⟩ _____

(2) $\begin{cases} 5x = 7y + 1 \\ 3y = 4x - 6 \end{cases}$

⟨Ans.⟩ _____

(3) $\begin{cases} 4x + 5y = -6 & \cdots ① \\ -2x - 3y = 2 & \cdots ② \end{cases}$

Rewrite ② : $x = \dfrac{\boxed{}}{2}$ $\cdots ③$

⟨Ans.⟩ _____

(4) $\begin{cases} 4y = 1 - 3x \\ -5x = -3 + 6y \end{cases}$

⟨Ans.⟩ _____

2 Use the following methods to solve the equations $\begin{cases} -3x-4y=-20 \\ 2x+5y=18 \end{cases}$.

15 points per question

(1) Addition/subtraction method

(2) Substitution method

⟨Ans.⟩ _____

⟨Ans.⟩ _____

3 Use the following methods to solve the equations $\begin{cases} 8x-y=-13 \\ 5x+3y=10 \end{cases}$.

15 points per question

(1) Addition/subtraction method

(2) Substitution method

⟨Ans.⟩ _____

⟨Ans.⟩ _____

You can solve equations with a few methods now!

© Kumon Publishing Co., Ltd.

Simultaneous Linear Equations

Date   /   /

Name

Score

/100

■ The Answer Key is on page 95.

1 Use the following methods to solve the equations $\begin{cases} 5x + 3y = 5 \\ 4x + y = -3 \end{cases}$.

10 points per question

(1)   Addition/subtraction method

(2)   Substitution method

⟨**Ans.**⟩ _____

⟨**Ans.**⟩ _____

2 Use the following methods to solve the equations $\begin{cases} x - 5y = 8 \\ 7x + 3y = -20 \end{cases}$.

10 points per question

(1)   Addition/subtraction method

(2)   Substitution method

⟨**Ans.**⟩ _____

⟨**Ans.**⟩ _____

3 **Use the following methods to solve the equations** $\begin{cases} 3x - 4y = -2 \\ 2x - 5y = 8 \end{cases}$.

15 points per question

(1) Addition/subtraction method

(2) Substitution method

⟨**Ans.**⟩ _____

⟨**Ans.**⟩ _____

4 **Use the following methods to solve the equations** $\begin{cases} 3x - 4y = -7 \\ -5x + 6y = 12 \end{cases}$.

15 points per question

(1) Addition/subtraction method

(2) Substitution method

⟨**Ans.**⟩ _____

⟨**Ans.**⟩ _____

When you have completed each exercise, compare the methods and think about which method was more efficient.

You're good at every method!

■ The Answer Key is on page 95.

1 **Write each word problem as equations, and then solve.**   20 points per question

(1) There are two numbers: x and y. The sum of the numbers is 15. The sum of 3 times x and 2 times y is 34. What are the values of x and y?

$$\begin{cases} x+y=15 \\ 3x+\boxed{}y=\boxed{} \end{cases}$$

⟨**Ans.**⟩ $x=\boxed{}$
            $y=\boxed{}$

(2) There are two numbers: a and b. The difference of the numbers is 3. The sum of 4 times a and 5 times b is 39. What are the values of a and b? (Assume that a is larger than b.)

⟨**Ans.**⟩ $a=\boxed{}$
            $b=\boxed{}$

(3) There are two numbers: f and g. 2 times the sum of f and g is 20. $\frac{2}{3}$ of f minus $\frac{1}{4}$ of g is 3. What are the values of f and g?

$$\begin{cases} 2(f+g)=\boxed{} \\ \frac{2}{3}f-\boxed{}g=3 \end{cases}$$

⟨**Ans.**⟩ $f=\boxed{}$
            $g=\boxed{}$

2 **Write each word problem as equations, and then solve.**

(1) If we fill a wheelbarrow with dirt, it weighs 60 pounds. When we only fill the wheelbarrow $\frac{1}{4}$ full, it weighs 33 pounds. How much does the empty wheelbarrow weigh?

Let x be the weight of the empty wheelbarrow
and y be the weight of the dirt in a full wheelbarrow.

$$\begin{cases} x+y=\boxed{} \\ x+\boxed{}\,y=33 \end{cases}$$

⟨Ans.⟩ _____

(2) If we fill a cart $\frac{1}{2}$ full with books, it weighs 50 pounds. When we fill a cart only $\frac{1}{3}$ full, it weighs 40 pounds. How much does the empty cart weigh?

Let x be the weight of the empty cart
and y be the weight of the books in a full cart.

⟨Ans.⟩ _____

Hooray for you!

Word Problems with Simultaneous Linear Equations

41

Level ★★★

Score

Date ___/___/___

Name

/100

■ The Answer Key is on page 95.

1 **Write the word problem as equations, and then solve.**

50 points for completion

Angela had a bracelet that was 3 parts gold and 1 part silver. Christina had another bracelet that was 1 part gold and 2 parts silver. Angela and Christina melted and combined both bracelets into a new necklace. The new necklace contains 6 ounces of gold and 7 ounces of silver. How many ounces did Angela's bracelet weigh, and how many ounces did Christina's bracelet weigh?

If Angela's bracelet had 3 parts gold and 1 part silver, then there were a total of $3+1=$ ☐ parts.

If Christina's bracelet had 1 part gold and 2 parts silver, then there were a total of ☐ parts .

If x represents the total weight of Angela's bracelet,

then $\dfrac{3}{\boxed{}}x$ represents the weight of the gold in Angela's bracelet

and ☐ represents the weight of the silver in Angela's bracelet.

If y represents the total weight of Christina's bracelet,

then ☐ represents the weight of the gold in Christina's bracelet

and ☐ represents the weight of the silver in Christina's bracelet.

$$\begin{cases} \dfrac{3}{\boxed{}}x + \dfrac{1}{3}y = 6 \\ \boxed{}x + \boxed{}y = 7 \end{cases}$$

⟨Ans.⟩ Angela's bracelet weighed _____ ounces.

Christina's bracelet weighed _____ ounces.

© Kumon Publishing Co., Ltd.

Suzette had a ring that was 1 part steel and 5 parts iron. Chayce had a ring that was 5 parts steel and 3 parts iron. Suzette and Chayce melted and combined both rings into a new ring. The new ring contains 7 grams of steel and 13 grams of iron. How many grams did Suzette's ring weigh, and how many grams did Chayce's ring weigh?

⟨**Ans.**⟩ Suzette's ring weighed _____ grams.

Chayce's ring weighed _____ grams.

You thought that through! Good job!

Date   /   /

Name

Score

 /100

■ The Answer Key is on page 96.

1 Determine the value of each expression when $a = 3$, $b = -\dfrac{1}{4}$, and $c = -2$.

5 points per question

(1) $a(b-c) =$

(3) $-c^3 - b^2 =$

(2) $(a-b) \div (b+c) =$

(4) $\dfrac{a^3}{bc^4} =$

2 Simplify each expression.

5 points per question

(1) $2x^2 + 3y - 6x^2 =$

(4) $2(a-b) - 3(-3a-b) =$

(2) $(-2x + y - 3z) + (x - 5y + 2z)$

=

(5) $\dfrac{3x-7}{5} - \dfrac{2x+9}{5} =$

(3) $\left(f - \dfrac{3}{2}g\right) - \left(-2g + \dfrac{1}{4}h\right)$

=

(6) $\dfrac{2x-3}{4} - \dfrac{-9x+7}{6} =$

3 Solve each equation.

5 points per question

(1) $3x + 2 = 14$

(3) $7x = -x - 10$

(2) $2x + 4 = \dfrac{1}{2}$

(4) $\dfrac{1}{2}x = -3 + \dfrac{2}{3}x$

4 Solve each equation, and check your answer.

10 points per question

(1) $2x + 6 = -x - 18$ 　　　　　Check!

(2) $\dfrac{3}{4}x - 5 = \dfrac{3}{2}x - 11$ 　　　　　Check!

5 Answer the word problem.

10 points for completion

Tony has 26 marbles, and Lisa has 7 marbles. If Tony gives Lisa x marbles, he will have 3 more marbles than Lisa. How many marbles did Tony give Lisa?

You're almost at the finish line!

⟨Ans.⟩ _____

Review

Date / /

Name

Level
★★★

Score
/100

■ The Answer Key is on page 96.

1 Solve the equations by using either the addition or subtraction method.

10 points per question

(1) $\begin{cases} 2x - 3y = -8 \\ 6x + 3y = 0 \end{cases}$

(2) $\begin{cases} 2x - 5y = -2 \\ 2x + 3y = 14 \end{cases}$

⟨Ans.⟩ _____

⟨Ans.⟩ _____

2 Solve the equations by using the substitution method.

10 points per question

(1) $\begin{cases} 3x + y = 11 \\ y = -2x + 8 \end{cases}$

(2) $\begin{cases} x = y - 2 \\ 2x - 3y = -9 \end{cases}$

⟨Ans.⟩ _____

⟨Ans.⟩ _____

3 **Solve each equation.**

10 points per question

(1) $\begin{cases} 2x - y = -5 \\ 3x + 2y = -4 \end{cases}$

(3) $\begin{cases} 2x + 5y = -4 \\ 3x - 4y = 17 \end{cases}$

⟨Ans.⟩ _____

⟨Ans.⟩ _____

(2) $\begin{cases} -3x + 7y = -8 \\ 6x + 5y = -3 \end{cases}$

(4) $\begin{cases} 9x + 2y = -12 \\ -6x - 5y = 19 \end{cases}$

⟨Ans.⟩ _____

⟨Ans.⟩ _____

4 **Answer the word problem.**

20 points for completion

Paul fills a jar with sand. When the jar is completely filled, it weighs 100 grams. When the jar is $\frac{3}{5}$ full of sand, it weighs only 76 grams. How much does the empty jar weigh?

⟨Ans.⟩ _____

Congratulations on completing
Algebra Workbook I!

Algebra
Workbook 1

---

① Fractions Review pp 2,3

① (1) $\frac{11}{18}$ (6) $\frac{3}{40}$

(2) $\frac{7}{20}$ (7) 1

(3) $1\frac{3}{20}$ (8) $3\frac{5}{8}$

(4) $1\frac{7}{12}$ (9) $8\frac{7}{24}$

(5) $2\frac{17}{40}$ (10) $9\frac{1}{8}$

② (1) $\frac{1}{32}$ (6) $1\frac{1}{2}$

(2) $1\frac{3}{8}$ (7) $\frac{18}{25}$

(3) $\frac{2}{5}$ (8) $\frac{2}{7}$

(4) $1\frac{3}{4}$ (9) $\frac{1}{5}$

(5) $2\frac{1}{4}$ (10) 6

---

② Exponents Review pp 4,5

① (1) 8   (6) $2^{\boxed{5}}=32$

(2) $1\frac{11}{25}$ (7) $\frac{16}{81}$

(3) $\frac{1}{36}$ (8) $\frac{9}{64}$

(4) $\frac{24}{25}$ (9) $\frac{64}{81}$

(5) $60\frac{3}{4}$ (10) $3\frac{3}{5}$

② (1) $2^{\boxed{4}}=16$ (6) $\frac{1}{36}$

(2) 243   (7) $\frac{27}{125}$

(3) $\frac{16}{81}$ (8) $\frac{1}{16}$

(4) 125   (9) 64

(5) $\frac{27}{64}$ (10) $2\frac{1}{4}$

---

③ Order of Operations Review pp 6,7

① (1) $6\frac{1}{2}$ (5) $7\frac{3}{4}$

(2) $1\frac{1}{4}$ (6) $13\frac{1}{2}$

(3) $\frac{3}{8}$ (7) $8\frac{2}{3}$

(4) 10   (8) $5\frac{1}{6}$

② (1) $1\frac{7}{12}$ (4) $2\frac{5}{6}$

(2) $1\frac{24}{25}$ (5) $9\frac{71}{72}$

(3) $\frac{24}{29}$ (6) $\frac{19}{72}$

---

④ Negative Numbers Review pp 8,9

① (1) -4 (8) $-2\frac{3}{20}$

(2) 6   (9) $-1\frac{1}{2}$

(3) 13   (10) $-3\frac{1}{2}$

(4) $-\frac{3}{4}$ (11) $1\frac{1}{2}$

(5) $6\frac{3}{5}$ (12) $4\frac{5}{6}$

(6) $-\frac{2}{3}$ (13) $-9\frac{5}{6}$

(7) $-2\frac{11}{12}$ (14) $-4\frac{5}{6}$

② (1) -12 (7) $-\frac{1}{40}$

(2) 32   (8) $-1\frac{1}{3}$

(3) $-\frac{3}{10}$ (9) $4\frac{1}{2}$

(4) -6 (10) $-1\frac{1}{2}$

(5) -6 (11) $\frac{3}{4}$

(6) 2   (12) $-2\frac{7}{10}$

---

⑤ Negative Numbers Review pp 10,11

① (1) -32 (6) $-\frac{1}{4}$

(2) -108 (7) $\frac{1}{4}$

(3) 36   (8) 108

(4) $\frac{1}{8}$ (9) $-\frac{40}{81}$

(5) -4 (10) $-8\frac{1}{3}$

② (1) $-\frac{2}{3}$ (6) $-1\frac{1}{8}$

(2) $-\frac{11}{14}$ (7) $-\frac{9}{32}$

(3) -48 (8) $30\frac{1}{9}$

(4) $-\frac{3}{40}$ (9) $-\frac{2}{5}$

(5) $-\frac{7}{15}$ (10) $\frac{16}{25}$

---

⑥ Values of Algebraic Expressions Review pp 12,13

① (1) -12 (6) -3

(2) $2\frac{1}{2}$ (7) 24

(3) -6 (8) $5\frac{1}{2}$

(4) 10   (9) $23\frac{1}{6}$

(5) -21 (10) $-3\frac{3}{4}$

② (1) $3\frac{1}{3}$ (6) 2

(2) $1\frac{1}{12}$ (7) $\frac{5}{6}$

(3) $5\frac{5}{6}$ (8) $-4\frac{1}{6}$

(4) $\frac{1}{6}$ (9) $-\frac{7}{20}$

(5) -9 (10) $3\frac{1}{2}$

⑦ Values of Algebraic Expressions Review pp 14,15

①
(1) -32
(2) 8
(3) -1
(4) $\dfrac{3}{4}$
(5) $-1\dfrac{2}{5}$
(6) -5
(7) -5
(8) 24
(9) -36
(10) -36

②
(1) $2\dfrac{19}{20}$
(2) -7
(3) $-\dfrac{1}{24}$
(4) $13\dfrac{1}{3}$
(5) $-\dfrac{3}{16}$
(6) $\dfrac{15}{64}$
(7) 40
(8) -4
(9) $\dfrac{7}{8}$
(10) $\dfrac{7}{8}$

⑧ Values of Algebraic Expressions Review pp 16,17

①
(1) -4
(2) $-2\dfrac{1}{2}$
(3) $-1\dfrac{5}{12}$
(4) $-\dfrac{1}{2}$
(5) $5\dfrac{1}{4}$
(6) -7
(7) $-1\dfrac{2}{3}$
(8) -7

②
(1) -4
(2) -3
(3) -3
(4) $2\dfrac{7}{8}$
(5) 3
(6) 3

⑨ Values of Algebraic Expressions Review pp 18,19

①
(1) 9
(2) $\dfrac{2}{5}$
(3) $-\dfrac{5}{6}$
(4) $-\dfrac{5}{6}$
(5) $31\dfrac{1}{4}$
(6) $24\dfrac{1}{2}$
(7) 1
(8) $-1\dfrac{11}{16}$

②
(1) $-\dfrac{2}{3}$
(2) $2\dfrac{1}{2}$
(3) $\dfrac{1}{6}$
(4) $\dfrac{5}{6}$
(5) $-9\dfrac{2}{27}$
(6) $-2\dfrac{4}{9}$

⑩ Simplifying Algebraic Expressions pp 20,21

①
(1) $5\boxed{x}$
(2) $15y$
(3) $8d$
(4) $4x$
(5) $8z$
(6) $5b$
(7) $3h$
(8) x
(9) $3y$
(10) $22w$

②
(1) $14x$
(2) $3y$
(3) $5a+\boxed{4a}=9a$
(4) $-13h$
(5) $7a$
(6) $5j$
(7) $-5j$
(8) $\dfrac{2}{7}w$
(9) $\dfrac{5}{9}z$
(10) $-\dfrac{1}{2}f$
(11) $\dfrac{\boxed{6}}{3}x+\dfrac{1}{3}x=\dfrac{\boxed{7}}{3}x$
(12) $\dfrac{11}{4}y$
(13) $\dfrac{7}{3}b$
(14) m
(15) $\boxed{3}xy$

⑪ Simplifying Algebraic Expressions pp 22,23

①
(1) $\boxed{6}x+5$
(2) $14y-3$
(3) $2h-6$
(4) $\dfrac{23}{12}bc+\dfrac{2}{7}$
(5) $-\dfrac{5}{4}ab-\dfrac{5}{6}$
(6) $-7x+5$
(7) $3x-11$
(8) $\boxed{-4}a+\boxed{7}b$
(9) $a-\dfrac{4}{3}b$
(10) $5w+y$
(11) $-\dfrac{13}{6}a-x$
(12) $\dfrac{7}{3}fg+2gh$
(13) $\dfrac{19}{8}c-\dfrac{3}{10}h$

②
(1) $\boxed{7}x^2-8x$
(2) $3x^2-7x$
(3) $\dfrac{9}{4}y^2-\dfrac{3}{2}$
(4) $-\dfrac{7}{3}a^2+\dfrac{1}{2}bc$
(5) $-s+6v^3$
(6) $5n^2-\dfrac{10}{3}w$
(7) $\boxed{8}a+\boxed{10}b+\boxed{13}c$
(8) $-2x-\dfrac{17}{4}y+\dfrac{3}{2}$
(9) $10b-6c^2+\dfrac{5}{6}$
(10) $\boxed{9}x^2+\boxed{4}$
(11) $3y^2+12$
(12) $-\dfrac{11}{4}abc+\dfrac{11}{2}be$

(12) Simplifying Algebraic Expressions pp 24, 25

(1)
(1) $4x - 2y$

(2) $-2a - 8b$

(3) $g + 3h$

(4) $-2m + 6n$

(5) $-\dfrac{3}{2}a + \dfrac{7}{2}b$

(6) $-\dfrac{7}{6}k + \dfrac{3}{4}l$

(7) $-x + \dfrac{5}{2}$

(8) $-\dfrac{5}{6}s - \dfrac{17}{20}t$

(2)
(1) $2a - 5b + 10c$

(2) $3x - 5y + z$

(3) $2a + \dfrac{5}{4}c$

(4) $-\dfrac{11}{4}b + \dfrac{9}{4}c^2 - \dfrac{11}{12}de$

(5) $5a + 5b - 3c + 2$

(6) $\dfrac{1}{4}e - \dfrac{3}{2}f + \dfrac{3}{2}g - \dfrac{1}{6}h$

(13) Simplifying Algebraic Expressions pp 26, 27

(1)
(1) $x \boxed{-} y \boxed{+} z$

(2) $a \boxed{-} b \boxed{+} c$

(3) $-g \boxed{+} h \boxed{+} i$

(4) $-l \boxed{+} m \boxed{-} n$

(5) $x \boxed{-} y \boxed{-} z$

(6) $a \boxed{+} b \boxed{+} c$

(7) $g \boxed{-} h \boxed{-} 2i$

(8) $l \boxed{-} 3m \boxed{-} n$

(2)
(1) $5x + 3$

(2) $13x + 3$

(3) $13x - 3$

(4) $-8x + 7$

(5) $-2x - 7$

(6) $-8x - 7$

(7) $\dfrac{7}{2}a + \dfrac{5}{2}$

(8) $-\dfrac{7}{2}a - \dfrac{5}{2}$

(9) $a - \dfrac{5}{2}b$

(10) $\dfrac{3}{4}x - \dfrac{4}{3}y$

(14) Simplifying Algebraic Expressions pp 28, 29

(1)
(1) $2x + 2y$

(2) $-2x - 2y$

(3) $8x - 2y$

(4) $-2x + 2y$

(5) $-8x - 10y$

(6) $-2a - b$

(7) $-2a + b$

(8) $-4a + 3b$

(9) $2a - 3b$

(10) $-4a + b$

(2)
(1) $6a - 12b$

(2) $-2a + 6b$

(3) $2a + 12b$

(4) $-6a + 6b$

(5) $-x + 4y$

(6) $-\dfrac{3}{4}f + \dfrac{8}{3}g$

(7) $-\dfrac{14}{15}y - \dfrac{3}{20}z$

(8) $\dfrac{13}{12}m - \dfrac{19}{12}n$

(9) $-\dfrac{15}{4}a + \dfrac{29}{10}b$

(10) $\dfrac{22}{15}x + \dfrac{17}{8}y$

(15) Simplifying Algebraic Expressions pp 30, 31

(1)
(1) $4x + 6y$

(2) $6x - 4y$

(3) $6x - 5y$

(4) $-\dfrac{4}{7}x - 3y$

(5) $2b + 13c$

(6) $6x - 15y$

(7) $\boxed{5}t$

(8) $\boxed{2}a$

(2)
(1) $10a + 8b + 5c$

(2) $11a + 4b - 5c$

(3) $-7x - 2y + z$

(4) $-\dfrac{3}{2}x^2 - \dfrac{2}{15}x - \dfrac{11}{4}$

(5) $\boxed{0}$

(6) $13x + 3y + \boxed{8}z$

(7) $2x + 4y - 9z$

(8) $a - 3b + 8c$

(9) $\dfrac{1}{6}x^2 - \dfrac{1}{3}xy + \dfrac{9}{20}y^2$

(10) $\dfrac{7}{12}a^2 - \dfrac{4}{3}a + 5$

(16) Simplifying Algebraic Expressions pp 32, 33

(1)
(1) $cd + \boxed{ce}$

(2) $\boxed{xy} + \boxed{xz}$

(3) $\boxed{ab} - \boxed{ac}$

(4) $2x - 2y$

(5) $3s - 3t$

(6) $3x + \boxed{2}x + 2y = 5x + 2y$

(7) $x + 3y$

(8) $-6x + 6y$

(9) $-\dfrac{5}{2}a + \dfrac{2}{3}b$

(10) $\dfrac{5}{4}m + \dfrac{1}{3}n$

② (1) $6x - 12y + \boxed{6}x - \boxed{10}y = 12x - 22y$

(2) $4x + 2y$

(3) $8x - 11y$

(4) $\dfrac{7}{3}a - b$

(5) $-\dfrac{19}{6}a^2 - a$

(6) $10x^2 + 25$

(7) $14x + 8y$

(8) $-x - 13y$

(9) $-7g + 13h$

(10) $-\dfrac{15}{2}a + \dfrac{15}{2}b$

(11) $-\dfrac{23}{10}c + \dfrac{1}{12}d$

(12) $\dfrac{9}{2}x + \dfrac{13}{2}y - \dfrac{13}{6}z$

⑰ Simplifying Algebraic Expressions   pp 34, 35

① (1) $\dfrac{(\boxed{6x+9}) + (\boxed{x-6})}{4} = \dfrac{7x+3}{4}$

(2) $\dfrac{-3x+7}{5}$

(3) $\dfrac{(\boxed{5x+11}) - (\boxed{3x+4})}{3} = \dfrac{2x+7}{3}$

(4) $\dfrac{x+14}{6}$

(5) $\dfrac{-(\boxed{4x+7}) - (\boxed{3x-2})}{2} = \dfrac{-7x-5}{2}$

(6) $\dfrac{-14x+12}{7}$

(7) $\dfrac{-6x+5}{4}$

(8) $\dfrac{x-8}{8}$

② (1) $\boxed{4}, \boxed{3}, \dfrac{20x+27}{12}$

(2) $\dfrac{18x-11}{6}$

(3) $\dfrac{8x-11}{10}$

(4) $\boxed{3}, \boxed{2}, \dfrac{7x+19}{12}$

(5) $\dfrac{12x-7}{18}$

(6) $\dfrac{7x+1}{4}$

(7) $\dfrac{-10x+1}{6}$

(8) $\dfrac{x-5}{9}$

(9) $\dfrac{5x+2}{3}$

(10) $\dfrac{-x+2}{5}$

⑱ Solving Equations   pp 36, 37

① (1) $x = 7 + \boxed{2}$

$x = 9$

(2) $x = -8 + \boxed{5}$

$x = \boxed{-3}$

(3) $x = -3$

(4) $x = 8$

(5) $x = -7$

(6) $x = -6$

(7) $x = 15$

(8) $x = 3$

(9) $x = 6$

(10) $x = 0$

② (1) $x = 4$

(2) $x = -4$

(3) $x = -5$

(4) $x = 0$

(5) $x = 2.6$

(6) $x = 1.5$

(7) $x = 4.1$

(8) $x = -3.5$

(9) $x = -\dfrac{4}{3}$

(10) $x = \dfrac{9}{4}$

(11) $x = -\dfrac{13}{2}$

(12) $x = \dfrac{5}{6}$

(13) $x = -\dfrac{1}{20}$

(14) $x = \dfrac{13}{4}$

(15) $x = \dfrac{3}{4}$

(16) $x = \dfrac{1}{2}$

⑲ Solving Equations   pp 38, 39

① (1) $x = 30 \times \dfrac{1}{\boxed{5}}$

$x = 6$

(2) $x = 7$

(3) $x = 6 \times (-\boxed{1})$

$x = -6$

(4) $x = 18 \times \left(-\dfrac{1}{\boxed{3}}\right)$

$x = -6$

(5) $x = -\dfrac{7}{2}$

(6) $x = 12$

(7) $x = -10$

(8) $x = \dfrac{3}{4}$

(9) $x = -18$

(10) $x = \dfrac{9}{2}$

② (1) $x = -8$

(2) $x = 15$

(3) $x = 12$

(4) $x = \dfrac{15}{2}$

(5) $x = -\dfrac{28}{3}$

(6) $x = -\dfrac{3}{16}$

(7) $x = -\dfrac{3}{8}$

(8) $x = \dfrac{6}{5}$

(9) $x = \dfrac{33}{28}$

(10) $x = -\dfrac{5}{14}$

(20) Solving Equations

pp 40, 41

1 (1) $3x = 10 + \boxed{2}$

$3x = \boxed{12}$

$x = \boxed{12} \times \dfrac{1}{\boxed{3}}$

$x = 4$

(2) $x = -3$

(3) $x = \dfrac{3}{4}$

(4) $x = \dfrac{7}{3}$

(5) $x = -1$

(6) $\dfrac{3}{5}x = 2 + \boxed{4}$

$\dfrac{3}{5}x = \boxed{6}$

$x = \boxed{6} \times \dfrac{5}{\boxed{3}}$

$x = 10$

2 (1) $x = 2$

(2) $x = -\dfrac{1}{3}$

(3) $x = -\dfrac{5}{6}$

(4) $x = \dfrac{4}{5}$

(5) $x = -2$

(6) $x = -\dfrac{1}{6}$

(7) $x = -24$

(8) $x = \dfrac{12}{5}$

(21) Solving Equations

pp 42, 43

1 (1) $2x + 6x - \boxed{7} = 9$

$\boxed{8}x - \boxed{7} = 9$

$x = 2$

(2) $x = -4$

(3) $x = -6$

(4) $x = -3$

2 (1) $x = -\dfrac{1}{5}$

(2) $x = -\dfrac{2}{13}$

(3) $x = \dfrac{7}{3}$

(4) $x = -18$

(5) $x = -\dfrac{5}{6}$

(6) $x = \dfrac{8}{7}$

(22) Solving Equations

pp 44, 45

1 (1) $\boxed{-}$, $\boxed{+}$, $x = 4$

(2) $\boxed{-}$, $\boxed{-}$, $x = -3$

(3) $x = -1$

(4) $x = 2$

(5) $x = -1$

(6) $x = -2$

2 (1) $x = 3$

(2) $x = -4$

(3) $x = -\dfrac{9}{2}$

(4) $x = 2$

(5) $x = \dfrac{1}{2}$

(6) $x = \dfrac{2}{3}$

(7) $x = \dfrac{16}{3}$

(8) $x = -\dfrac{5}{8}$

(23) Solving Equations

pp 46, 47

1 (1) $2x + \boxed{3x} = 4 + \boxed{6}$

$x = 2$

Left side $= 2 \times \boxed{2} - 6 = -2$

Right side $= -3 \times \boxed{2} + 4 = -6 + 4 = -2$

(2) $x = -2$

Left side $= 4 \times (-2) + 7 = -1$

Right side $= 9 + 5 \times (-2) = -1$

2 (1) $x = -2$

(2) $x = 6$

(3) $x = \dfrac{4}{3}$

(4) $x = \dfrac{3}{25}$

(24) Solving Equations

pp 48, 49

1 (1) $2x - \boxed{10} = -4$

$x = 3$

(2) $x = -8$

(3) $x = 1$

(4) $x = 2$

(5) $x = 8$

(6) $x = \dfrac{9}{2}$

2 (1) $4 + \boxed{2}x + \boxed{8} = -x$

$x = -4$

(2) $x = -1$

(3) $x = -3$

(4) $x = -\dfrac{4}{3}$

(5) $x = \dfrac{8}{5}$

(6) $x = \dfrac{8}{3}$

(7) $x = -\dfrac{8}{15}$

(8) $x = -18$

(25) Solving Equations

pp 50, 51

1 (1) $\boxed{12}$, $\boxed{12}$, $\boxed{12}$, $x = -2$

(2) $x = 1$

(3) $x = -24$

(4) $x = -\dfrac{60}{11}$

2 (1) $\boxed{20}$, $\boxed{20}$, $x = 4$

(2) $x = 8$

(3) $x = -\dfrac{3}{4}$

(4) $x = -\dfrac{3}{11}$

(5) $x = -8$

(6) $x = -1$

(26) Word Problems with Equations pp 52, 53

1 (1) $3+x=\boxed{7}$, $x=4$ (6) $x=6\times(\boxed{-2})$, $x=-12$

(2) $\frac{1}{2}+x=-3$,   (7) $\frac{3}{4}\times x=6$, $x=8$

$x=-\frac{7}{2}$ $\left(\text{or}-3\frac{1}{2}\right)$

(3) $\boxed{x}-2=10$, $x=12$ (8) $\frac{x}{5}=4$, $x=20$

(4) $x-\frac{1}{2}=\frac{1}{3}$, $x=\frac{5}{6}$ (9) $\frac{x}{3}=\frac{2}{5}$, $x=\frac{6}{5}$

(5) $5-x=\boxed{4}$, $x=1$ (10) $\frac{x}{12}=\frac{2}{3}$, $x=8$

2 (1) $4\times x=\boxed{76}$, $x=19$ **Ans.** 19 yo-yos

(2) $24\times x=192$, $x=8$ **Ans.** 8 ounces

(3) $5\times3+\boxed{4}\times x=39$, $x=6$ **Ans.** 6 oranges

(4) $25\times3+40\times x=315$, $x=6$ **Ans.** 6 grams

(27) Word Problems with Equations pp 54, 55

1 (1) **Ans.** $(\boxed{12}-x)$ pens

(2) **Ans.** $(30-x)$ crayons

(3) **Ans.** $(x-2)$ cats

(4) **Ans.** $(25-x)$ girls

(5) **Ans.** $(5+x)$ girls

(6) **Ans.** $(x-5)$ boys

2 Method #1: $\boxed{6}$, $\boxed{6}$, $x=9$, $24-9=15$ **Ans.** 9 boys / 15 girls

Method #2: $\boxed{6}$, $\boxed{x-6}$, $x=15$, $24-15=9$ **Ans.** 9 boys / 15 girls

3 Method #1:
If x is the number of bananas,
then $(x+9)$ is the number of apples.
$x+(x+9)=31$
$x=11$
$31-11=20$ **Ans.** 20 apples / 11 bananas

Method #2:
If x is the number of apples,
then $(x-9)$ is the number of bananas.
$x+(x-9)=31$
$x=20$
$31-20=11$ **Ans.** 20 apples / 11 bananas

(28) Word Problems with Equations pp 56, 57

1 (1) $\boxed{x}$, $\boxed{8}$, $22-\boxed{x}=\boxed{8}+x$, $x=7$ **Ans.** 7 markers

(2) $\boxed{x}$, $20-\boxed{x}=(6+x)+\boxed{4}$, $x=5$ **Ans.** 5 coins

2 (1) $\boxed{x}$, $(12+\boxed{x})=\boxed{4}\times12$, $x=36$ **Ans.** 36 years

(2) $x+10=3\times x$, $x=5$ **Ans.** 5 years old

(2) $\boxed{x}$, $\boxed{x}$, $\boxed{2}\times(10+\boxed{x})=26+\boxed{x}$,
$x=6$ **Ans.** 6 years

(29) Solving Equations pp 58, 59

1 (1) $x=b-\boxed{6}$ (6) $x=a-b$

(2) $x=c-1$ (7) $x=-c-d$

(3) $x=a+\boxed{8}$ (8) $x=b+\boxed{1}$

(4) $x=d+4$ (9) $x=-y-2$

(5) $x=-g+11$ (10) $x=a-3$

2 (1) $x=\frac{c}{a}$ (6) $x=\frac{-c+d}{b}$

(2) $x=\frac{b}{\boxed{-3}}$ (7) $cx=d-\boxed{3}$

$x=-\frac{b}{\boxed{3}}$ $x=\frac{d-3}{c}$

(3) $x=\frac{g}{f}$ (8) $x=\frac{b+5}{a}$

(4) $4x=\boxed{s}$ (9) $rx=\boxed{s-2t}$

$x=\frac{s}{4}$ $x=\frac{s-2t}{r}$

(5) $x=\frac{7}{a}$ (10) $x=\frac{-y+2z}{w}$

(30) Solving Equations pp 60, 61

1 (1) $\boxed{ad}$, $x=\frac{abd-ac}{d}$ (4) $\boxed{bc}$, $x=\frac{cd}{b}$

(2) $x=-\frac{24}{11}$ (5) $x=\frac{bc}{ad}$

(3) $x=-\frac{10}{13}$ (6) $x=-\frac{15fh}{4g}$

2 (1) $x=a+b$ (5) $x=a+3b$

(2) $\boxed{2}$, $x=a+\frac{b}{2}$ (6) $x=-5d-3$

(3) $x=\frac{a}{y}-4$ (7) $x=-\frac{5a-3}{4}$

(4) $x=\frac{c+d}{a}-b$ (8) $x=-\frac{a(b-d)}{M}$

(31) Simultaneous Linear Equations — pp 62,63

(1) (1) 6, -1, -1, -12, -4

Ans. $(x, y)=(-1, -4)$

(2) 2 Ans. $(x, y)=(4, 5)$

(2) (1) Ans. $(x, y)=(-1, 2)$ (4) Ans. $(x, y)=(4, 5)$

(2) Ans. $(x, y)=(-6, 3)$ (5) Ans. $(x, y)=(-5, 2)$

(3) Ans. $(x, y)=(2, 1)$

(32) Simultaneous Linear Equations — pp 64,65

(1) (1) Ans. $(x, y)=(2, -1)$ (2) Ans. $(x, y)=(-5, -3)$

(2) (1) $+$ Ans. $(x, y)=(-4, -1)$

(2)   Ans. $(x, y)=\left(\dfrac{1}{2}, -4\right)$

(3)   Ans. $(x, y)=(6, 3)$

(4)   Ans. $(x, y)=\left(\dfrac{7}{10}, \dfrac{6}{5}\right)$

(33) Simultaneous Linear Equations — pp 66,67

(1) (1) 2, 22, 27, 3, 3, 5 Ans. $(x, y)=(3, 5)$

(2)   Ans. $(x, y)=(2, -1)$

(2) (1)   Ans. $(x, y)=(3, 2)$

① : $2\times 3 + 3\times 2 = 6 + 6 = 12$

② : $5\times 3 - 2 = 15 - 2 = 13$

(2)   Ans. $(x, y)=(1, -3)$

(34) Simultaneous Linear Equations — pp 68,69

(1) (1) ①$\times 4$: $\begin{cases} 20x+12y=-28 & \cdots③ \\ \end{cases}$

②$\times 3$: $\begin{cases} 18x+12y=-24 & \cdots④ \\ \end{cases}$

③$-$④ : $\qquad 2x=-4$

$\qquad\qquad x=-2$

$5\times(-2)+3y=-7$

$\qquad 3y=3$

$\qquad y=1$

Ans. $(x, y)=(-2, 1)$

(2) 2 Ans. $(x, y)=(4, 2)$

(2) (1) — Ans. $(x, y)=(6, 2)$

① : $3\times 6 - 5\times 2 = 18 - 10 = 8$

② : $-2\times 6 + 7\times 2 = -12 + 14 = 2$

(2)   Ans. $(x, y)=(-1, 0)$

(3)   Ans. $(x, y)=\left(-2, \dfrac{1}{3}\right)$

(35) Simultaneous Linear Equations — pp 70,71

(1) (1) -3, 6, 2, 7, -4 Ans. $(x, y)=(6, -4)$

(2)   Ans. $(x, y)=(5, 3)$

(3)   Ans. $(x, y)=(-2, -4)$

(4)   Ans. $(x, y)=(4, 9)$

(2) (1) 6, 4 Ans. $(x, y)=(4, -1)$

(2) 2, 3, 8 Ans. $(x, y)=(-4, 2)$

(3)   Ans. $(x, y)=(-1, 2)$

(4)   Ans. $(x, y)=\left(\dfrac{19}{6}, \dfrac{7}{3}\right)$

(36) Simultaneous Linear Equations — pp 72,73

(1) (1) 2, -6, 4 Ans. $(x, y)=(-1, 3)$

(2)   Ans. $(x, y)=(-2, 8)$

(3) 5, 5, 8, $\therefore -5x+2y=-21$

Ans. $(x, y)=(5, 2)$

(4)   Ans. $(x, y)=\left(\dfrac{10}{3}, \dfrac{2}{3}\right)$

(2) (1) ①$\times 12$: $4(x-y)+3(2x-y)=24$

②$\times 6$: $3(x-y)+2(x+y)=7$

$\therefore \begin{cases} 10x-7y=24 \\ 5x-y=7 \end{cases}$

Ans. $(x, y)=(1, -2)$

(2)   Ans. $(x, y)=(-1, 0)$

37 Simultaneous Linear Equations pp 74,75

1 (1) **Ans.** $(x, y)=(3, 7)$ (2) **Ans.** $(x, y)=(2, -2)$

2 (1) Rewrite ② : $x=\boxed{-3}y-1$ ···③

$4(\boxed{-3}y-1)+5y=10$

$\boxed{-7}y=14$

$y=\boxed{-2}$

$x=\boxed{6}-1$

Ans. $(x, y)=(\boxed{5}, \boxed{-2})$

(2) **Ans.** $(x, y)=(6, 6)$

(3) $\boxed{8-5x}$ **Ans.** $(x, y)=(1, 1)$

(4) **Ans.** $(x, y)=(-2, 4)$

38 Simultaneous Linear Equations pp 76,77

1 (1) Rewrite ① : $x=\dfrac{-3y+7}{\boxed{2}}$ ···③

$3\left(\dfrac{-3y+7}{\boxed{2}}\right)-4y=2$

$3(-3y+7)-8y=\boxed{4}$

$-17y=\boxed{-17}$

$y=\boxed{1}$

$x=\dfrac{\boxed{-3}+7}{2}$

Ans. $(x, y)=(2, 1)$

(2) **Ans.** $(x, y)=(3, 2)$

(3) $\boxed{-3y-2}$ **Ans.** $(x, y)=(-4, 2)$

(4) **Ans.** $(x, y)=(3, -2)$

2 (1) **Ans.** $(x, y)=(4, 2)$ (2) **Ans.** $(x, y)=(4, 2)$

3 (1) **Ans.** $(x, y)=(-1, 5)$ (2) **Ans.** $(x, y)=(-1, 5)$

39 Simultaneous Linear Equations pp 78,79

1 (1) **Ans.** $(x, y)=(-2, 5)$ (2) **Ans.** $(x, y)=(-2, 5)$

2 (1) **Ans.** $(x, y)=(-2, -2)$ (2) **Ans.** $(x, y)=(-2, -2)$

3 (1) **Ans.** $(x, y)=(-6, -4)$ (2) **Ans.** $(x, y)=(-6, -4)$

4 (1) **Ans.** $(x, y)=\left(-3, -\dfrac{1}{2}\right)$ (2) **Ans.** $(x, y)=\left(-3, -\dfrac{1}{2}\right)$

40 Word Problems with Simultaneous Linear Equations pp 80,81

1 (1) $\begin{cases} x+y=15 \\ 3x+\boxed{2}y=\boxed{34} \end{cases}$

Ans. $\begin{aligned} x&=\boxed{4} \\ y&=\boxed{11} \end{aligned}$

(2) $\begin{cases} a-b=3 \\ 4a+5b=39 \end{cases}$

Ans. $\begin{aligned} a&=\boxed{6} \\ b&=\boxed{3} \end{aligned}$

(3) $\begin{cases} 2(f+g)=\boxed{20} \\ \dfrac{2}{3}f-\dfrac{1}{4}g=3 \end{cases}$

Ans. $\begin{aligned} f&=\boxed{6} \\ g&=\boxed{4} \end{aligned}$

2 (1) $\begin{cases} x+y=\boxed{60} \\ x+\dfrac{1}{\boxed{4}}y=33 \end{cases}$

Ans. 24 pounds

(2) $\begin{cases} x+\dfrac{1}{2}y=50 \\ x+\dfrac{1}{3}y=40 \end{cases}$

Ans. 20 pounds

41 Word Problems with Simultaneous Linear Equations pp 82,83

1 $\boxed{4}$, $\boxed{3}$, $\boxed{\dfrac{3}{4}}$, $\boxed{\dfrac{1}{4}x}$, $\boxed{\dfrac{1}{3}y}$, $\boxed{\dfrac{2}{3}y}$

$\begin{cases} \dfrac{\boxed{3}}{\boxed{4}}x+\dfrac{1}{3}y=6 \\ \dfrac{\boxed{1}}{\boxed{4}}x+\dfrac{\boxed{2}}{\boxed{3}}y=7 \end{cases}$

Ans. Angela's bracelet weighed 4 ounces.
Christina's bracelet weighed 9 ounces.

2 If x represents the total weight of Suzette's ring, and y represents the total weight of Chayce's ring, the equation is the following.

$\begin{cases} \dfrac{1}{6}x+\dfrac{5}{8}y=7 \\ \dfrac{5}{6}x+\dfrac{3}{8}y=13 \end{cases}$

Ans. Suzette's ring weighed 12 grams.
Chayce's ring weighed 8 grams.

1 (1) $5\frac{1}{4}$ (3) $7\frac{15}{16}$

(2) $-1\frac{4}{9}$ (4) $-6\frac{3}{4}$

2 (1) $-4x^2+3y$ (4) $11a+b$

(2) $-x-4y-z$ (5) $\frac{x-16}{5}$

(3) $f+\frac{1}{2}g-\frac{1}{4}h$ (6) $\frac{24x-23}{12}$

3 (1) $x=4$ (3) $x=-\frac{5}{4}$

(2) $x=-\frac{7}{4}$ (4) $x=18$

4 (1) $x=-8$

(2) $x=8$

5 $(26-x)=(7+x)+3, \; x=8$ **Ans.** 8 marbles

1 (1) **Ans.** $(x, y)=(-1, 2)$ (2) **Ans.** $(x, y)=(4, 2)$

2 (1) **Ans.** $(x, y)=(3, 2)$ (2) **Ans.** $(x, y)=(3, 5)$

3 (1) **Ans.** $(x, y)=(-2, 1)$ (3) **Ans.** $(x, y)=(3, -2)$

(2) **Ans.** $(x, y)=\left(\frac{1}{3}, -1\right)$ (4) **Ans.** $(x, y)=\left(-\frac{2}{3}, -3\right)$

4 Let x be the weight of the jar and y be
the weight of sand need to fill the jar.

$$\begin{cases} x+y=100 \\ x+\frac{3}{5}y=76 \end{cases}$$

$(x, y)=(40, 60)$

Ans. 40 grams

KUMON MATH WORKBOOKS

Algebra
Workbook II

Table of Contents

KUMON

Solving Equations Review

Date / /   Name

■ The Answer Key is on page 184.

 1   **Solve each equation.**

5 points per question

Examples

$$x - 4 = 5$$
$$x = 5 + 4$$
$$x = 9$$

$$3 + x = 10$$
$$x = 10 - 3$$
$$x = 7$$

To solve equations, think in terms of "x = a number." For questions such as the examples above, add or subtract a number on both sides of the equal sign.

(1) $x - 3 = 8$

(2) $1 + x = 7$

(3) $x - 5 = -4$

(4) $3 + x = -5$

(5) $-2 + x = -6$

(6) $x + 1 = \dfrac{1}{4}$

(7) $x - \dfrac{1}{3} = 2$

(8) $\dfrac{1}{4} + x = \dfrac{1}{3}$

(9) $\dfrac{2}{3} + x = -3$

(10) $-\dfrac{5}{2} + x = -\dfrac{2}{3}$

To check if your answer is correct, substitute your value for x into the original equation.

2 **Solve each equation.**

Examples

$$3x = 12$$
$$x = 12 \times \frac{1}{3}$$
$$x = 4$$

$$-\frac{1}{4}x = 7$$
$$x = 7 \times (-4)$$
$$x = -28$$

For questions such as the examples above, multiply both sides of the equal sign by a number.

(1) $2x = 10$

(2) $\frac{1}{3}x = -8$

(3) $-9x = 12$

(4) $6x = -6$

(5) $-\frac{1}{5}x = -2$

(6) $\frac{2}{3}x = 5$

(7) $-\frac{4}{3}x = 6$

(8) $5x = -\frac{1}{3}$

(9) $-\frac{3}{4}x = -\frac{1}{2}$

(10) $-\frac{5}{2}x = -\frac{5}{4}$

You're off to a great start!

Solving Equations Review

Date / /   Name

Score /100

■ The Answer Key is on page 184.

1   **Solve each equation.**

6 points per question

Example

$$6x - 11 = 4x - 3$$
$$6x - 4x = -3 + 11$$
$$2x = 8$$
$$x = 4$$

Transposition (or **Inverse Operations**) is when we move a term to the other side of the equation and change the sign.

(1) $7x + 8 = 5x + 14$

(2) $8x + 4 = 5x + 10$

(3) $10x + 1 = 9x - 3$

(4) $3x - 3 = 5x + 7$

(5) $2x - 1 = -2x + 5$

(6) $-6x - 2 = -x + 12$

(7) $\frac{1}{2}x + 6 = -\frac{3}{2}x - 1$

(8) $-\frac{1}{3}x + \frac{1}{2} = -\frac{2}{5}x + \frac{1}{4}$

(2) Solve each equation, and check your answer.

13 points per question

Example

$$7x - 6 = 4x + 9$$
$$7x - 4x = 9 + 6$$
$$3x = 15$$
$$x = 5$$

Check !

Left side $= 7 \times 5 - 6 = 35 - 6 = 29$

Right side $= 4 \times 5 + 9 = 20 + 9 = 29$

If the value on the left side of the equal sign matches the value on the right, then your answer is correct.

(1) $3x - 6 = x + 8$

Check!

Left side $=$

Right side $=$

(2) $2x + 15 = 9x - 6$

Check!

Left side $=$

Right side $=$

(3) $-\dfrac{1}{3}x + \dfrac{1}{2} = \dfrac{5}{3}x - \dfrac{3}{2}$

Check!

Left side $=$

Right side $=$

(4) $-\dfrac{1}{2}x - \dfrac{1}{3} = -\dfrac{3}{4}x + \dfrac{2}{3}$

Check!

Left side $=$

Right side $=$

Get in the habit of double checking your answers!

101

Date   /   /   Name

■ The Answer Key is on page 184.

1 Solve each equation.

6 points per question

Examples

$$2(3x+4)=-4$$

$$6x+8=-4$$

$$6x=-12$$

$$x=-2$$

$$\frac{1}{2}x+\frac{1}{6}=\frac{1}{3}$$

$$\left(\frac{1}{2}x+\frac{1}{6}\right)\times 6=\left(\frac{1}{3}\right)\times 6$$

$$3x+1=2$$

$$3x=1$$

$$x=\frac{1}{3}$$

To simplify an equation, remove the parentheses and denominators.

(1) $3(x+2)=-9$

(4) $\frac{1}{4}x+\frac{1}{3}=\frac{1}{6}$

To remove denominators, multiply each side by the least common multiple (LCM) of the denominators.

(2) $3(3x-2)=2(2x+2)$

(5) $\frac{2}{3}x-1=-\frac{1}{8}+\frac{3}{4}x$

(3) $-(x-1)=-4(2-x)$

2 **Solve each equation for** x.

Examples
$$x+2=a \qquad\qquad a=b-x \qquad\qquad 2x=a-b$$
$$x=a-2 \qquad\qquad x=-a+b \qquad\qquad x=\frac{a-b}{2}$$

Transpose as necessary to formulate each question as "$x=$ a number."

(1) $x+3=y$

$x=y-3$

(6) $4x=-b$

(2) $c=-b-x$

(7) $-ax=-b$

(3) $x-y=-w$

(8) $9x=4a+b$

(4) $-3=-f-x$

(9) $ax=-2b+3c$

(5) $w=-8-x$

(10) $rx-s=4t$

You're acing this review section!

Simultaneous Linear Equations Review

Date	Name	Score
/ /		/100

■ The Answer Key is on page 184.

 1 **Solve each equation for both variables by using the subtraction method.**

10 points per question

Example

$$\begin{cases} 4x+5y=7 & \cdots ① \\ x+5y=13 & \cdots ② \end{cases}$$

$$① - ②: \quad \begin{array}{r} 4x+5y=7 \quad \cdots ① \\ -) \ x+5y=13 \quad \cdots ② \\ \hline 3x \qquad =-6 \\ x \qquad =-2 \end{array}$$

Substitute this into ② :

$$-2+5y=13$$
$$5y=15$$
$$y=3$$

⟨Ans.⟩ $(x, y)=(-2, 3)$

Subtraction method

Step 1: Number each equation.

Step 2: Remove a variable by subtracting one equation from the other.

Step 3: Substitute the value of the variable into either original equation (usually the simpler one) to solve for the other variable.

Step 4: Write your answer.

(1) $\begin{cases} 7x+4y=1 \\ 5x+4y=3 \end{cases}$

(3) $\begin{cases} -4x+3y=-14 \\ -4x-2y=6 \end{cases}$

⟨Ans.⟩ _____

⟨Ans.⟩ _____

(2) $\begin{cases} 2x-6y=10 \\ 2x+y=-11 \end{cases}$

(4) $\begin{cases} -3x+6y=5 \\ x+6y=-7 \end{cases}$

⟨Ans.⟩ _____

⟨Ans.⟩ _____

2 Solve each equation for both variables by using the addition method.

15 points per question

Example

$$\begin{cases} 3x + 2y = 12 & \cdots ① \\ 5x - 2y = 4 & \cdots ② \end{cases}$$

① + ② :
$$\begin{array}{r} 3x + 2y = 12 \quad \cdots ① \\ +)\ 5x - 2y = 4 \quad \cdots ② \\ \hline 8x \quad\quad = 16 \\ x \quad\quad = 2 \end{array}$$

Substitute this into ② :
$$10 - 2y = 4$$
$$-2y = -6$$
$$y = 3$$

⟨Ans.⟩ $(x, y) = (2, 3)$

Addition method

Step 1: Number each equation.

Step 2: Remove a variable by adding one equation to the other.

Step 3: Substitute the value of the variable into either original equation (usually the simpler one) to solve for the other variable.

Step 4: Write your answer.

(1) $\begin{cases} 4x - 3y = 2 \\ -x + 3y = -5 \end{cases}$

⟨Ans.⟩ _____

(2) $\begin{cases} 5x - 2y = -9 \\ -5x - y = 0 \end{cases}$

⟨Ans.⟩ _____

(3) $\begin{cases} x + 3y = 2 \\ \dfrac{1}{4}x - 3y = 3 \end{cases}$

⟨Ans.⟩ _____

(4) $\begin{cases} -\dfrac{1}{2}x + 2y = -1 \\ \dfrac{1}{2}x - 5y = 7 \end{cases}$

⟨Ans.⟩ _____

You know this well!

© Kumon Publishing Co., Ltd.   105

Simultaneous Linear Equations Review

Date   /   /

Name

■ The Answer Key is on page 184.

1 **Solve each equation.**

20 points per question

Example

$$\begin{cases} 3x - 5y = 2 & \cdots ① \\ x + 2y = 8 & \cdots ② \end{cases}$$

Step 1: Number each equation.

② × 3 :　　　$3x + 6y = 24$　$\cdots ③$

Step 2: Multiply one equation by a constant and number the new equation. In this example, we multiplied equation ② by 3 and called it equation ③.

① − ③ :　　　$3x - 5y = 2$　$\cdots ①$

　　　　　$-\,) \; 3x + 6y = 24 \;\cdots ③$

　　　　　　　$-11y = -22$

　　　　　　　　$y = 2$

Step 3: Remove a variable by using the addition or subtraction method.

Substitute this into ② :

　　　$x + 4 = 8$

　　　$x = 4$

⟨**Ans.**⟩ $(x, y) = (4, 2)$

Step 4: Substitute the value of the variable into either original equation (usually the simpler one) to solve for the other variable.

Step 5: Write your answer.

(1) $\begin{cases} 6x + 5y = -3 \\ 3x - 2y = 12 \end{cases}$

(2) $\begin{cases} 5x - 2y = 8 \\ 3x + 8y = 14 \end{cases}$

⟨**Ans.**⟩ _____

⟨**Ans.**⟩ _____

2 Solve each equation, and check your answer.

30 points per question

(1) $\begin{cases} x - 2y = -8 & \cdots ① \\ 2x + 3y = -2 & \cdots ② \end{cases}$

Check!

① : $\boxed{} - 2 \times \boxed{} =$

The left side of ① matches the right!

② : $2 \times \boxed{} + 3 \times \boxed{} =$

The left side of ② matches the right!

⟨Ans.⟩ _____

(2) $\begin{cases} -10x - 9y = -1 \\ 4x - 3y = -4 \end{cases}$

Check!

⟨Ans.⟩ _____

 Check your answer by substituting the values for x and y into both original equations to be sure that both equations are satisfied.

Checking your work takes time, but it's worth it.

■ The Answer Key is on page 184.

1 Solve each equation.

20 points per question

Example

$$\begin{cases} 5x+2y=4 & \cdots ① \\ 7x+3y=5 & \cdots ② \end{cases}$$

Step 1: Number each equation.

①×3 : $15x+6y=12$ $\cdots ③$

②×2 : $14x+6y=10$ $\cdots ④$

Step 2: Multiply one equation by a constant and number the new equation.

Step 3: Multiply the other equation by a constant and number the new equation.

③−④ : $15x+6y=12$ $\cdots ③$

$$-)\ 14x+6y=10\ \cdots ④$$

$$\overline{\hspace{1.5cm} x \hspace{1cm} =2}$$

Step 4: Remove a variable by using the addition or subtraction method.

Substitute this into ① :

$$10+2y=4$$

$$2y=-6$$

$$y=-3$$

Step 5: Substitute the value of the variable into either original equation (usually the simpler one) to solve for the other variable.

⟨**Ans.**⟩ $(x, y)=(2, -3)$

Step 6: Write your answer.

(1) $\begin{cases} 7x-3y=20 & \cdots ① \\ 5x+4y=2 & \cdots ② \end{cases}$

(2) $\begin{cases} 3x+8y=1 & \cdots ① \\ 5x+6y=9 & \cdots ② \end{cases}$

①×3 :

②×☐ :

⟨**Ans.**⟩ _____

⟨**Ans.**⟩ _____

2 **Solve each equation, and check your answer.**

20 points per question

(1) $\begin{cases} 3x - 7y = -2 \\ 5x - 4y = 12 \end{cases}$

Check!

⟨**Ans.**⟩ _____

(2) $\begin{cases} 3x + 4y = 24 \\ 5x - 6y = 2 \end{cases}$

Check!

⟨**Ans.**⟩ _____

(3) $\begin{cases} 6x + 5y = 10 \\ 9x + 8y = 13 \end{cases}$

Check!

⟨**Ans.**⟩ _____

Remember you have to eliminate one variable in order to solve for the other variable.

Bravo!

Level ☆

Score
/100

■ The Answer Key is on page 185.

1 **Solve each equation.**

10 points per question

Example

$$\begin{cases} y = 2x - 3 & \cdots ① \\ 7x - 4y = 8 & \cdots ② \end{cases}$$

Substitute ① into ② :

$$7x - 4(2x - 3) = 8$$
$$7x - 8x + 12 = 8$$
$$-x = -4$$
$$x = 4$$

Substitute this into ① :

$$y = 8 - 3$$
$$y = 5$$

⟨Ans.⟩ $(x, y) = (4, 5)$

Substitution method

Step 1: Number each equation.

Step 2: Substitute one equation into the other to remove a variable.

Step 3: Substitute the value of the variable into either original equation (usually the simpler one) to solve for the other variable.

Step 4: Write your answer.

(1) $\begin{cases} y = 3x + 1 \\ x - 2y = 8 \end{cases}$

(3) $\begin{cases} x = -5 - 3y \\ -x - y = -1 \end{cases}$

⟨Ans.⟩ _____

⟨Ans.⟩ _____

(2) $\begin{cases} x = y - 5 \\ 2x - 5y = -4 \end{cases}$

(4) $\begin{cases} 4x - 7y = -3 \\ y = 2 - x \end{cases}$

⟨Ans.⟩ _____

⟨Ans.⟩ _____

2 Solve each equation by using the substitution method. 15 points per question

(1) $\begin{cases} 2x + y = -2 & \cdots \text{①} \\ 3x + 2y = -1 & \cdots \text{②} \end{cases}$

       Rewrite ① : $y = \boxed{} x - 2 \cdots \text{③}$

       Substitute ③ into ② :

       ⟨Ans.⟩ _____

(3) $\begin{cases} 4x + 3y = -1 & \cdots \text{①} \\ 6x + 5y = -3 & \cdots \text{②} \end{cases}$

       Rewrite ① : $y = \dfrac{\boxed{}}{3} \cdots \text{③}$

       Substitute ③ into ② :

       ⟨Ans.⟩ _____

(2) $\begin{cases} 2x - 5y = 5 \\ x + 3y = 8 \end{cases}$

       ⟨Ans.⟩ _____

(4) $\begin{cases} 5x - 8y = 7 \\ -2x + 7y = 1 \end{cases}$

       ⟨Ans.⟩ _____

Terrific attention to detail!

Simultaneous Linear Equations Review

Date   /   /

Name

■ The Answer Key is on page 185.

1 **Use the following methods to solve the equations** $\begin{cases} x + 3y = -8 & \cdots① \\ 4x - 3y = 13 & \cdots② \end{cases}$ · 10 points per question

(1)   Addition/subtraction method

(2)   Substitution method

⟨**Ans.**⟩ _____

⟨**Ans.**⟩ _____

2 **Use the following methods to solve the equations** $\begin{cases} 4x - 3y = 7 & \cdots① \\ 3x + y = -11 & \cdots② \end{cases}$ · 10 points per question

(1)   Addition/subtraction method

(2)   Substitution method

⟨**Ans.**⟩ _____

⟨**Ans.**⟩ _____

3 **Use the following methods to solve the equations** $\begin{cases} 2x - 5y = 5 & \cdots ① \\ -3x + 4y = 3 & \cdots ② \end{cases}$.

15 points per question

(1)   Addition/subtraction method

(2)   Substitution method

〈Ans.〉

〈Ans.〉

4 **Use the following methods to solve the equations** $\begin{cases} 5x + 6y = -16 & \cdots ① \\ -7x - 4y = -4 & \cdots ② \end{cases}$.

15 points per question

(1)   Addition/subtraction method

(2)   Substitution method

〈Ans.〉

〈Ans.〉

You can do every method!

Date / /

Name

■ The Answer Key is on page 185.

1 **Solve each equation.**

25 points per question

(1) $\begin{cases} x+y+z=6 & \cdots① \\ x-2y+3z=1 & \cdots② \\ x+3y-z=8 & \cdots③ \end{cases}$

Eliminate x:

①$-$②: $\begin{cases} 3y-2z=\boxed{} & \cdots④ \\ -2y+\boxed{}z=\boxed{} & \cdots⑤ \end{cases}$
①$-$③:

④$+$⑤: $\quad y=\boxed{}$

Substitute this into ④:

$3\times\boxed{}-2z=\boxed{}$

$-2z=\boxed{}$

$z=\boxed{}$

Substitute $y=\boxed{}$ and $z=\boxed{}$ into ①:

$x+\boxed{}+\boxed{}=6$

$x=\boxed{}$

⟨**Ans.**⟩ $(x,y,z)=(\boxed{},\boxed{},\boxed{})$

Don't forget!

Three linear equations with the same variables are called **Simultaneous Linear Equations in Three Variables**.

To solve, eliminate one variable so there are only two linear equations with two variables. Then you can use the addition, subtraction, or substitution method to solve.

(2) $\begin{cases} -x+2y+z=1 & \cdots① \\ 2x-5y-z=2 & \cdots② \\ 3x-7y-z=3 & \cdots③ \end{cases}$

Eliminate z:

①$+$②:

②$-$③:

Think, "What is the easiest variable to eliminate?"

⟨**Ans.**⟩

2 **Solve each equation.**

25 points per question

(1) $\begin{cases} x - 2y - 4z = 4 & \cdots ① \\ -2x - 2y - 3z = -7 & \cdots ② \\ 3x + 2y - 6z = 2 & \cdots ③ \end{cases}$

Eliminate y:

$① - ②$: $\begin{cases} 3x - z = \boxed{} & \cdots ④ \\ x - \boxed{}z = \boxed{} & \cdots ⑤ \end{cases}$
$② + ③$:

$⑤ \times 3$:

⟨Ans.⟩ _____

(2) $\begin{cases} 3x + y + 2z = 1 & \cdots ① \\ -5x - 3y - 2z = 5 & \cdots ② \\ -9x - 6y - 2z = 13 & \cdots ③ \end{cases}$

⟨Ans.⟩ _____

Wow! You can solve for three variables!

Level

Score

/100

Date / /

Name

■ The Answer Key is on page 185.

1 **Solve each equation.**

25 points per question

(1) $\begin{cases} 2x - 3y - 4z = -5 & \cdots ① \\ x - 5y + 3z = 4 & \cdots ② \\ -x + 2y + 3z = 5 & \cdots ③ \end{cases}$

Eliminate x:

② × 2 : $2x - 10y + 6z = 8$ ⋯④

① − ④ :

② + ③ :

⟨Ans.⟩ _____

(2) $\begin{cases} 2x + 3y + 2z = -5 & \cdots ① \\ 5x + 6y + 6z = -7 & \cdots ② \\ 6x - 3y + 7z = 1 & \cdots ③ \end{cases}$

Eliminate ☐ :

Which is the easiest variable to eliminate?

⟨Ans.⟩ _____

2 **Solve each equation.**

25 points per question

(1) $\begin{cases} 4x - 3y - 3z = -5 & \cdots ① \\ -x + 2y + z = 2 & \cdots ② \\ -3x + 5y + 2z = 7 & \cdots ③ \end{cases}$

Eliminate z :

② × 3 : $-3x + 6y + \boxed{}z = \boxed{}$ $\cdots ④$

① + ④ : $\cdots ⑤$

② × 2 : $\cdots ⑥$

③ − ⑥ :

⟨**Ans.**⟩

(2) $\begin{cases} 3x - 2y - 7z = -7 & \cdots ① \\ 4x - 4y - 3z = 7 & \cdots ② \\ 5x + y - 8z = 8 & \cdots ③ \end{cases}$

Eliminate $\boxed{}$:

⟨**Ans.**⟩

Do your best!

Simultaneous Linear Equations in Three Variables

Date   /   /

Name

Score

/100

■ The Answer Key is on page 185.

1 **Solve each equation.**

50 points for completion

$$\begin{cases} 5x + 3y - 7z = -23 & \cdots ① \\ 3x + 5y - 5z = -9 & \cdots ② \\ 2x + 7y - 3z = 2 & \cdots ③ \end{cases}$$

Eliminate x :

① × 3 : $\boxed{}x + 9y - 21z = -69$ $\cdots ④$

② × $\boxed{}$: $15x + \boxed{}y - \boxed{}z = \boxed{}$ $\cdots ⑤$

④ − ⑤ : $\boxed{}y + 4z = \boxed{}$ $\cdots ⑥$

Simplify ⑥ :

$\boxed{}y + z = \boxed{}$ $\cdots ⑦$

② × 2 :

③ × 3 :

⟨Ans.⟩ _____

(2) **Solve each equation.**

50 points for completion

$$\begin{cases} -5x + 3y + 10z = 8 & \cdots ① \\ -7x + 5y + 8z = 2 & \cdots ② \\ -8x + 6y + 3z = -5 & \cdots ③ \end{cases}$$

⟨Ans.⟩

Your hard work really shows!

Word Problems with Simultaneous Linear Equations

Level ★

Date	Name	Score
/ /		/100

■ The Answer Key is on page 185.

1 **Answer each word problem.**

25 points per question

(1) Donald's class goes to a play, and the students must sit on benches. If 6 students sit on each bench, then 7 students will remain standing. If 8 students sit on each bench, then 1 student will remain standing. How many benches and how many students are there?

Let x be the number of benches and y be the number of students:

$$\begin{cases} 6x + 7 = y \\ \boxed{}x + 1 = y \end{cases}$$

⟨**Ans.**⟩ ☐ benches

☐ students

(2) Irene's class takes a train back to school. If there are 3 students in each train car, then 16 must wait for the next train. If there are 5 students in each train car, then no students must wait. How many train cars and how many students are there?

⟨**Ans.**⟩

2 Answer each word problem.

(1) Jeff's teacher has a bag of apples and oranges, where there are 3 times as many oranges as apples. If each student receives 2 apples, there is 1 apple left over. If each student receives 5 oranges, there are 7 oranges left over. How many students and apples are there?

Let x be the number of students and y be the number of apples:

$\boxed{}\,x + 1 = y \cdots ①$

Because y is the number of apples, and there are 3 times as many oranges as apples, there are $\boxed{}\,y$ oranges.

$5x + 7 = \boxed{}\,y \cdots ②$

Substitute ① into ② :

Substitute this into ① :

⟨Ans.⟩
_____

(2) Ying's teacher has a box of pens and crayons, and there are 2 times as many crayons as pens. If each student receives 2 pens, there are 6 pens left over. If each student receives 5 crayons, there are 7 crayons left over. How many students and pens are there?

⟨Ans.⟩
_____

You are better than ever!

Word Problems with Simultaneous Linear Equations

Level ☆

13

Date　　　/　　　/

Name

Score

/100

■ The Answer Key is on page 185.

1 **Answer each word problem.**　　　25 points per question

(1)　Warren rows 5 miles per hour. How many hours will it take him to row 20 miles?

Let x be the number of hours,

$5x = \boxed{}$

⟨Ans.⟩ _____

(2)　Viola normally rows 10 miles per hour, but the wind is blowing against her at 2 miles per hour. How many hours will it take her to row 40 miles?

Viola normally rows 10 miles per hour,

but the wind is blowing against her at 2 miles per hour,

therefore Viola's speed is: $\left(10 - \boxed{}\right) = \boxed{}$ miles per hour.

Let x be the number of hours,

$\boxed{} x = 40$

⟨Ans.⟩ _____

(3)　Rob normally rows 12 miles per hour, and the wind is pushing him along at 3 miles per hour. How many hours will it take him to row 45 miles?

⟨Ans.⟩ _____

© Kumon Publishing Co., Ltd.

2 Answer the word problem.

25 points for completion

Jane took 5 hours to row 40 miles against the current. She then turned around and took only 2 hours to row 60 miles with the current. How fast does Jane normally row (without any current), and how fast was the current?

Let x be Jane's normal speed and y be the speed of the current.

Because she took 5 hours to row 40 miles against the current:

$$\left(x - \boxed{}\right) \times 5 = 40$$

Because she took 2 hours to row 60 miles with the current:

$$\left(x + \boxed{}\right) \times 2 = 60$$

⟨**Ans.**⟩ Jane's speed:

The current's speed:

You show great dedication!

Inequalities

Date   /   /

Name

Level ☆

Score   /100

■ The Answer Key is on page 185.

1 **Solve each equation.**   2 points per question

Don't forget!

An **inequality** is a mathematical sentence that has a symbol such as $>$ or $<$. Solving an inequality is similar to solving an equation.

Examples

$x + 3 > 5$

$x > 5 - 3$

$x > 2$

In the example to the left, the "$+3$" was transposed to the other side.

$2x > 14$

$x > 7$

In the example to the left, both sides were divided by 2.

The symbol $>$ is read as "is greater than," and the symbol $<$ is read as "is less than."

(1) $x + 2 > 8$

(2) $x + 5 > 2$

(3) $x - 6 > -2$

(4) $2x > 10$

(5) $3x > 12$

(6) $2x > 7$

2 **Solve each inequality.**   2 points per question

(1) $2x + 8 > 20$

$2x > 12$

$x > 6$

(2) $2x - 6 > 16$

(3) $3x + 9 > 12$

(4) $6x + 7 > 7$

3 Solve each inequality. 5 points per question

(1) $x - 3 < 7$

(2) $x - 4 < 9$

(3) $x + 8 < 6$

(4) $2x < 18$

(5) $5x < 13$

(6) $2x + 8 < 5$
$\qquad 2x < -3$

(7) $3x + 2 < -10$

(8) $4x - 8 < 6$

(9) $9x + 9 < 8$

4 Solve each inequality. 5 points per question

When multiplying or dividing an inequality by a negative number, reverse the inequality symbol.

Examples $\quad -2x < 6 \qquad\qquad\qquad\qquad -3x > -15$
$\qquad\qquad\qquad x > -3 \qquad\qquad\qquad\qquad\quad x < 5$

(1) $-2x < 10$
$\quad x \boxed{} -5$

(2) $-x > -3$
$\quad x \boxed{} 3$

(3) $-4x < 8$

(4) $-6x < -12$

(5) $-2x + 2 > -6$
$\quad -2x > \boxed{}$

(6) $-6x - 8 > 4$

(7) $-x + 9 < -5$

You've got this!

Inequalities

15

Date / /

Name

Level ☆

Score

/100

■ The Answer Key is on page 186.

1 **Solve each inequality.** 5 points per question

> Inequalities with ≥ or ≤ are solved the same way as inequalities with < or >.
>
> **Examples**
> $$x + 2 \geq 6$$
> $$x \geq 4$$
>
> $$-3x - 8 \leq 7$$
> $$-3x \leq 15$$
> $$x \geq -5$$
>
> The symbol ≥ is read as "is greater than or equal to," and the symbol ≤ is read as "is less than or equal to."

(1) $2x + 5 \geq 7$

(5) $-2x - (x - 3) \geq -6$

(2) $-3x + 6 \leq 1$

(6) $-3x - (4 - 7x) \geq -10 + x$

(3) $5x - 2(x - 3) \leq 18$
    $5x - \boxed{}x + \boxed{} \leq 18$

(7) $-3x - \left(\dfrac{1}{2} - x\right) \leq -2$

(4) $6x + 2(x + 4) \leq 8$

(8) $-5x - \left(\dfrac{3}{4}x - 1\right) \geq 3$

Don't forget to reverse the inequality symbol when multiplying or dividing an inequality by a negative number!

2 Solve each inequality.

Example

$$\frac{x-5}{3} > \frac{x-2}{4}$$

Multiply both sides by the LCM of the denominators, 12:

$$4(x-5) > 3(x-2)$$

$$4x-20 > 3x-6$$

$$x > 14$$

(1) $\dfrac{x+4}{2} < \dfrac{x-7}{3}$

(4) $\dfrac{5x-1}{6} \geq \dfrac{3x+1}{4}$

(2) $\dfrac{2x-1}{5} \geq \dfrac{4x+5}{7}$

(5) $\dfrac{x+9}{3} \geq \dfrac{4x-5}{9}$

(3) $3x-5 \leq \dfrac{3-x}{4}$

(6) $\dfrac{3x-2}{4} < \dfrac{9x+8}{10}$

 Think of solving an inequality like solving an equation.

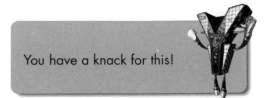 You have a knack for this!

16

Word Problems with Inequalities

Level ☆

Date / /

Name

Score /100

■ The Answer Key is on page 186.

1 **Write each word problem as an inequality, and then solve.**

8 points per question

(1) x plus 6 is greater than four times x.

$$x + 6 > \boxed{} x$$

⟨**Ans.**⟩ _____

(2) y minus 8 is less than or equal to 6.

⟨**Ans.**⟩ _____

(3) 5 times z is greater than or equal to z plus 6.

⟨**Ans.**⟩ _____

(4) -2 multiplied by w is less than or equal to 3 less than w.

⟨**Ans.**⟩ _____

(5) The sum of 3 and q is greater than 8 times q.

⟨**Ans.**⟩ _____

2 **Write each word problem as an inequality, and then solve.**

15 points per question

(1) 3 times x plus 2 is less than 10.

$\boxed{}\,x+2<\boxed{}$

⟨**Ans.**⟩ _____

(2) 3 times the sum of x plus 2 is less than 10.

$\boxed{}\,(x+\boxed{})<10$

⟨**Ans.**⟩ _____

(3) 2 times the sum of $3x$ and 5 is greater than or equal to 9 times x.

⟨**Ans.**⟩ _____

(4) One half the difference of x and 4 is less than twice the difference of 5 and x.

$\dfrac{1}{2}\,(x-\boxed{})<2\,(\boxed{}-x)$

⟨**Ans.**⟩ _____

Hats off to you!

Score

/ 100

Date   /   /

Name

■ The Answer Key is on page 186.

1 **Draw each inequality on a number line.**   10 points per question

Example $x > 2$

The open dot (○) shows that 2 is not included in the solution.

(1) $x < 4$

(3) $x > 0$

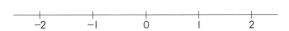

(2) $x > -1$

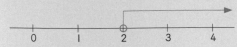

(4) $x > -\dfrac{1}{2}$

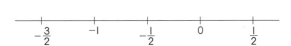

2 **Draw each inequality on a number line.**   10 points per question

Example $x \le -1$

A closed dot (●) shows that −1 is included in the solution.

(1) $x \ge 0$

(2) $x \le -\dfrac{5}{2}$

3 **Draw each inequality on a number line to find the range of x that satisfies both inequalities.**

10 points per question

Examples

$x > 1$ ···①
$x < 3$ ···②

⟨**Ans.**⟩ $1 < x < 3$

$x < 0$ ···①
$x > 1$ ···②

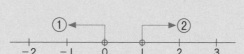

⟨**Ans.**⟩ No solution

The shaded region shows the range of x that satisfies both inequalities.

(1) $\begin{cases} x < 5 & ···① \\ x > 2 & ···② \end{cases}$

⟨**Ans.**⟩ ☐ $< x <$ ☐

(2) $\begin{cases} x > 6 & ···① \\ x < 0 & ···② \end{cases}$

⟨**Ans.**⟩ No solution

(3) $\begin{cases} x \geq -1 & ···① \\ x < 0 & ···② \end{cases}$

⟨**Ans.**⟩ ☐ $\leq x <$ ☐

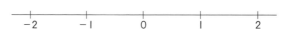

(4) $\begin{cases} x \leq -7 & ···① \\ x \geq -10 & ···② \end{cases}$

⟨**Ans.**⟩ ☐ $\leq x \leq$ ☐

Nicely done!

Date   /   /

Name

Level
★★

Score
/100

■ The Answer Key is on page 186.

1 Draw each inequality on a number line to find the range of x that satisfies both inequalities.

4 points per question

Examples

$x > 1$ ···①
$x > 3$ ···②
⟨Ans.⟩ $x > 3$

$x > -2$ ···①
$x \geq 0$ ···②
⟨Ans.⟩ $x \geq 0$

(1) $\begin{cases} x < 3 & \cdots① \\ x < -1 & \cdots② \end{cases}$

⟨Ans.⟩ _____

(2) $\begin{cases} x \geq 3 & \cdots① \\ x > -3 & \cdots② \end{cases}$

⟨Ans.⟩ _____

(3) $\begin{cases} x \leq 4 & \cdots① \\ x < 0 & \cdots② \end{cases}$

⟨Ans.⟩ _____

(4) $\begin{cases} x \geq 2 & \cdots① \\ x \geq -1 & \cdots② \end{cases}$

⟨Ans.⟩ _____

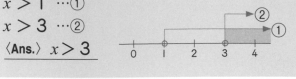

(5) $\begin{cases} x \leq -4 & \cdots① \\ x < -3 & \cdots② \end{cases}$

⟨Ans.⟩ _____

© Kumon Publishing Co., Ltd.

2 **Find the range of x that satisfies both inequalities.**

(1) $\begin{cases} x > 2 \\ x < 8 \end{cases}$

⟨Ans.⟩ _____

(2) $\begin{cases} x > 0 \\ x < -1 \end{cases}$

⟨Ans.⟩ _____

(3) $\begin{cases} x \geq 2 \\ x \geq -1 \end{cases}$

⟨Ans.⟩ _____

(4) $\begin{cases} x \leq -2 \\ x > -1 \end{cases}$

⟨Ans.⟩ _____

(5) $\begin{cases} x \leq 5 \\ x < -6 \end{cases}$

⟨Ans.⟩ _____

(6) $\begin{cases} 3x < 2x + 8 \quad \cdots ① \\ 2x \geq 6 \qquad \cdots ② \end{cases}$

① becomes: $x < \boxed{}$

② becomes: $x \geq \boxed{}$

⟨Ans.⟩ $\boxed{} \leq x < \boxed{}$

(7) $\begin{cases} x > 4x - 6 \qquad \cdots ① \\ 5x - 3 > 7x + 9 \quad \cdots ② \end{cases}$

① becomes:

② becomes:

⟨Ans.⟩ _____

(8) $\begin{cases} 6(x - 1) \leq -2(x + 3) \quad \cdots ① \\ 2 - 6x < -2 + 2x \qquad \cdots ② \end{cases}$

① becomes:

② becomes:

⟨Ans.⟩ _____

You really know your stuff!

■ The Answer Key is on page 187.

1 **Answer each word problem.**

25 points per question

(1) Jack has 50 coins, and Tracey has 20 coins. They each give James the same amount of coins. Jack's new amount of coins is less than or equal to 4 times Tracey's new amount of coins. Find the range of the amount of coins they each could have given James.

Let x be the amount of coins they each gave James.

Jack's new amount of coins is: $50 - x$.

Tracey's new amount of coins is: $20 - \boxed{}$,

therefore $50 - x \leq \boxed{} \left(20 - \boxed{}\right)$

⟨**Ans.**⟩ _____ coins or fewer

(2) Maria has 60 toys, and Alex has 40 toys. They each give Jane the same amount of toys. Alex's new amount of toys is greater than or equal to $\frac{1}{2}$ of Maria's new amount of toys. Find the range of the amount of toys they each could have given Jane.

⟨**Ans.**⟩ _____

2 Answer each word problem.

25 points per question

(1) Many boys and girls are playing in a park. There are twice as many girls as there are boys. If there are more than 60 children in the park, find the range of the amount of boys that could be in the park.

Let x be the number of boys.

Because there are twice as many girls as boys,

the number of girls is ☐x.

Because there are more than 60 children,

$x + $☐$x > 60$

⟨**Ans.**⟩ more than   boys

(2) Bruce sells TVs and radios at his store. The number of radios is half of the number of TVs. If there are at most 285 items in his store, find the range of the amount of radios.

⟨**Ans.**⟩

This is tough. You're doing great!

Graphs

20

Date   /   /

Name

Level ☆

Score

/100

■ The Answer Key is on page 187.

1 Write the *x*-coordinate or *y*-coordinate of each point.   4 points per question

Examples

Point A: The *x*-coordinate is 1. The *y*-coordinate is 4.

Point B: The *x*-coordinate is −1. The *y*-coordinate is 3.

- The *x*-**axis** is the horizontal number line, and the *y*-**axis** is the vertical number line.
- Each point is expressed as a pair of numbers called the *x*-**coordinate** and the *y*-**coordinate**.

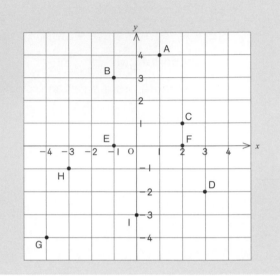

(1) Point C: The *x*-coordinate is 2. The *y*-coordinate is ☐.

(2) Point D: The *x*-coordinate is 3. The *y*-coordinate is ☐.

(3) Point E: The *x*-coordinate is ☐. The *y*-coordinate is 0.

(4) Point F: The *x*-coordinate is 2. The *y*-coordinate is ☐.

(5) Point G: The *x*-coordinate is ☐. The *y*-coordinate is −4.

(6) Point H: The *x*-coordinate is ☐. The *y*-coordinate is ☐.

(7) Point I: The *x*-coordinate is 0. The *y*-coordinate is ☐.

2 Write the coordinates of each point.

6 points per question

Don't forget!

A point is expressed as a pair of coordinates or ordered pair. Write the x-coordinate and the y-coordinate inside the parentheses.

Example

The coordinates of point A are: $(3, -1)$

(1) Point B: $\left(3, \boxed{}\right)$

(2) Point C: $\left(\boxed{}, -\dfrac{1}{2}\right)$

(3) Point D: $\left(-\dfrac{3}{2}, \boxed{}\right)$

(4) Point E: $\left(\boxed{}, \boxed{}\right)$

(5) Point F: $\left(0, \boxed{}\right)$

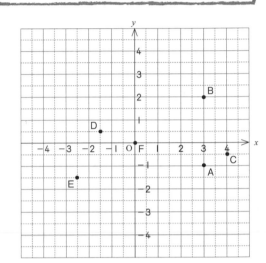

 Point F is located at the **origin**. The origin is where the x-axis and y-axis intersect.

3 Answer each question.

7 points per question

(1) What are the coordinates of point A? (,)

(2) What are the coordinates of point B? (,)

(3) What are the coordinates of point C? (,)

(4) What are the coordinates of point D? (,)

(5) What are the coordinates of point E? (,)

(6) Which point is located at the origin? $\boxed{}$

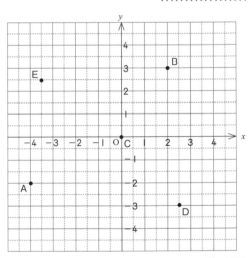

You are on point!

137

21 Graphs

Date / /

Name

Level
☆

Score
/100

■ The Answer Key is on page 187.

1 **Plot each point on the coordinate plane.**

5 points per question

Examples

Point A: $(-2, 0)$

Point B: $\left(\dfrac{1}{2}, 3\right)$

(1) Point C: $(2, 6)$

(2) Point D: $(0, 0)$

(3) Point E: $(2, -1)$

(4) Point F: $\left(-\dfrac{7}{2}, -6\right)$

(5) Point G: $(-3, 4)$

(6) Point H: $(-5, 0)$

(7) Point I: $(0, -4)$

(8) Point J: $\left(\dfrac{9}{2}, 2\right)$

(9) Point K: $\left(3, -\dfrac{5}{2}\right)$

(10) Point L: $\left(-\dfrac{7}{2}, -\dfrac{3}{2}\right)$

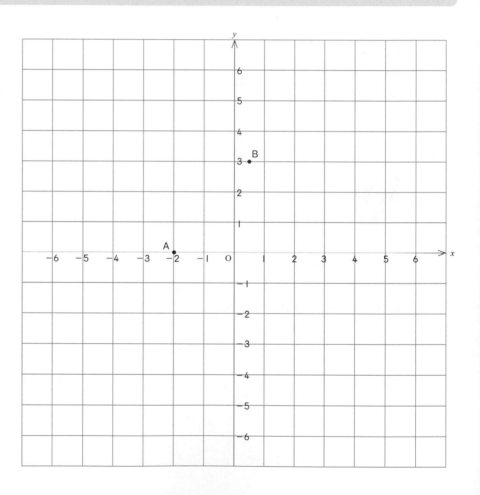

A point is expressed as a pair of coordinates in the form of: (x, y).

2 **Plot each point on the coordinate plane.**

5 points per question

(1) Point A: $(-3, 1)$

(2) Point B: $(0, -2)$

(3) Point C: $\left(-1, -\dfrac{3}{2}\right)$

(4) Point D: $\left(6, \dfrac{5}{2}\right)$

(5) Point E: $\left(-\dfrac{1}{2}, -\dfrac{9}{2}\right)$

(6) Point F: $\left(\dfrac{7}{2}, 4\right)$

(7) Point G: $(3, -6)$

(8) Point H: $(5, 0)$

(9) Point ☐ lies on the x-axis

(10) Point ☐ lies on the y-axis

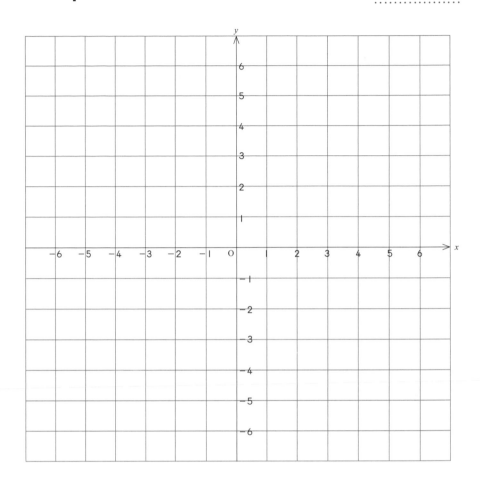

Well done!

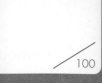

■ The Answer Key is on page 187.

1 **Find each value of y for the equation $y = 2x - 3$ according to each value of x.**

4 points per question

(1) $x = -3$ $y = 2 \times (-3) - 3 =$

(2) $x = -2$ $y =$

(3) $x = -1$ $y =$

(4) $x = 0$ $y =$

(5) $x = 1$ $y =$

(6) $x = 2$ $y =$

(7) $x = 3$ $y =$

(8) $x = \dfrac{1}{2}$ $y =$

(9) $x = \dfrac{3}{4}$ $y =$

(10) $x = -\dfrac{3}{2}$ $y =$

2 **Graph the equation** $y = x - 2$ **by completing each exercise.**

(1) Find each value of y according to each value of x.

5 points per question

① A : $x = -2$ $y =$

③ C : $x = 1$ $y =$

② B : $x = 0$ $y =$

④ D : $x = -\dfrac{1}{2}$ $y =$

(2) Plot each of the above points on the coordinate plane, and then draw a line to connect the points.

20 points for completion

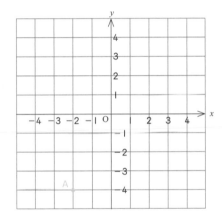

The line represents the graph of $y = x - 2$.

3 **Find each value of y for the equation** $y = -2x + 1$ **according to each value of x. Then graph the equation.**

20 points for completion

x	-1	$-\dfrac{1}{2}$	0	$\dfrac{1}{2}$	1	2
y						

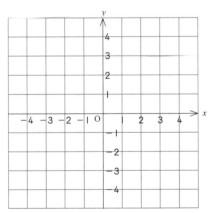

Nicely done!

■ The Answer Key is on page 188.

1 **Daniel walks at a speed of 3 miles per hour. Use this information to complete each exercise.**

10 points per question

(1) Let x represent the number of hours that Daniel walks, and let y represent the distance. Complete the chart to show the relationship between x and y.

x (hour)	0	1	2	2.5	3	4
y (mile)	0	3				

(2) Express the relationship between x and y as an equation.

$$y = \boxed{}\, x$$

(3) Graph the equation.

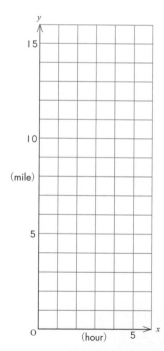

(4) How many miles has Daniel walked after 5 hours? Use the graph to answer.

⟨Ans.⟩ _____

┌ Don't forget! ────────────────────────

A **direct relation** is when the value of y increases as the value of x increases (and vice versa).

© Kumon Publishing Co., Ltd.

2 Jacob rows a boat at a speed of 12 miles per hour. Use this information to complete each exercise.

(1) Let x represent the number of hours that Jacob rows, and let y represent the distance. Complete the chart to show the relationship between x and y.

x (hour)	1	2	5	8	10
y (mile)					

(2) Express the relationship between x and y as an equation.

(3) Graph the equation.

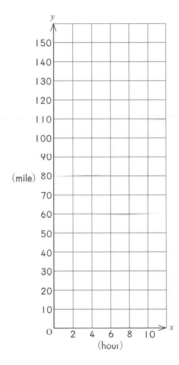

(4) How many miles has Jacob rowed after 7 hours? Use the equation to answer.

$y = 12 \times \boxed{} =$

⟨Ans.⟩ _____

You're moving fast!

24

Word Problems with Graphs

Level ☆☆

Score
/100

Date / /

Name

■ The Answer Key is on page 188.

1 Matt is practicing his jump shot. The chart below shows the relationship between the amount of time he practices (*x* minutes) and the number of baskets he makes (*y* baskets). Use the chart to complete each exercise.

12 points per question

x (minutes)	5	10	15	20
y (baskets)	10	20	30	40

(1) Express the relationship between x and y as an equation.

⟨Ans.⟩ _____

(2) Graph the equation.

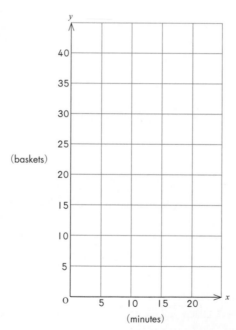

(3) How many baskets has Matt made after 18 minutes?

$y =$

⟨Ans.⟩ _____

2 Felix has a machine that makes gumballs. The chart below shows the relationship between the amount of time the machine works (x minutes) and the number of gumballs it makes (y gumballs). Use the chart to complete each exercise.

16 points per question

x (minutes)	2	3	5	8
y (gumballs)	16	24	40	64

(1) Express the relationship between x and y as an equation.

⟨Ans.⟩ _____

(2) Graph the equation.

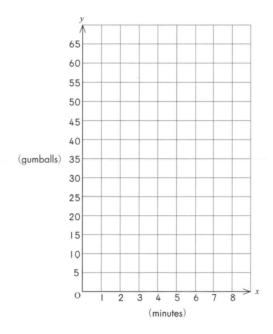

(3) How many gumballs are made after 15 minutes? Use the equation to answer.

⟨Ans.⟩ _____

(4) How many gumballs are made after 9.5 minutes? Use the equation to answer.

⟨Ans.⟩ _____

Way to go!

Graphs
y-Intercept

Date   /   /

Name

Level ★★

Score   /100

■ The Answer Key is on page 188.

1 Determine the *y*-intercept of each line.   8 points per question

The **y-intercept** of a line is the *y*-coordinate of the point where the line crosses the *y*-axis.

Examples

The *y*-intercept of line (1) is −2.
The *y*-intercept of lines (2) and (3) is 3.

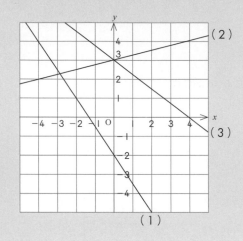

(1) The *y*-intercept of line (1) is ☐.

(2) The *y*-intercept of line (2) is ☐.

(3) The *y*-intercept of line (3) is ☐.

(4) The *y*-intercept of line (4) is ☐.

(5) The *y*-intercept of line (5) is ☐.

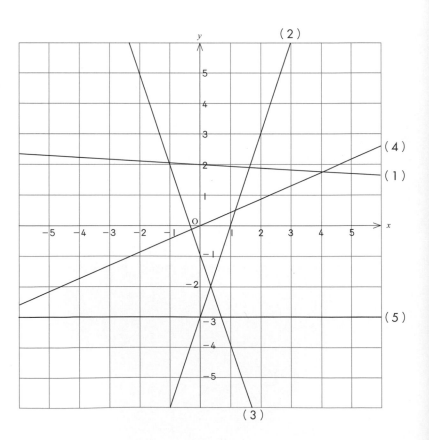

2 **Graph each equation to find the y-intercept.**

7 points per question

(1) The y-intercept of $y = 2x + 1$ is ☐.

(2) The y-intercept of $y = x - 3$ is ☐.

(3) The y-intercept of $y = \dfrac{3}{2}x - 1$ is ☐.

(4) The y-intercept of $y = \dfrac{1}{2}x + 4$ is ☐.

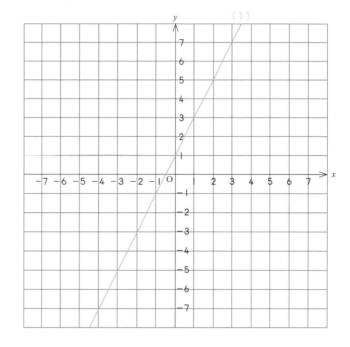

Label each line with the question number.

3 **Determine the y-intercept of each equation.**

8 points per question

Examples

The y-intercept of $y = 2x - 5$ is -5.

The y-intercept of $y = \dfrac{1}{2}x + 3$ is 3.

(1) The y-intercept of $y = 4x - 1$ is ☐.

(3) The y-intercept of $y = \dfrac{1}{2}x - \dfrac{3}{5}$ is ☐.

(2) The y-intercept of $y = x + 2$ is ☐.

(4) The y-intercept of $y = \dfrac{5}{3}x + \dfrac{16}{7}$ is ☐.

If you need to, make a table with the values of x and y.

You are doing really well!

Graphs
Slope

Date   /   /

Name

Level

Score

/ 100

■ The Answer Key is on page 188.

1 **Determine the slope of each line.**   8 points per question

Don't forget!

The **slope** of a line describes the steepness of a line.

$$\text{slope} = \frac{rise}{run}$$

For example, the slope of line (1) is $\frac{3}{1}$ or 3 because the line rises 3 units vertically for every 1 unit it runs horizontally. The slope of line (2) is $\frac{1}{2}$ because the line rises 1 unit vertically for every 2 units it runs horizontally.

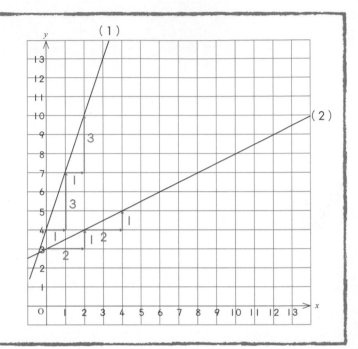

(1)   The slope of line (1) is ☐.

(2)   The slope of line (2) is ☐.

(3)   The slope of line (3) is ☐.

(4)   The slope of line (4) is ☐.

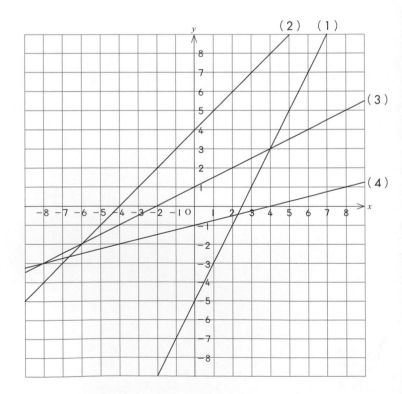

© Kumon Publishing Co., Ltd.

2 **Graph each equation to find the slope.**

7 points per question

(1) The slope of $y = 3x$ is ☐.

(2) The slope of $y = x$ is ☐.

(3) The slope of $y = \dfrac{1}{2}x + 2$ is ☐.

(4) The slope of $y = \dfrac{2}{3}x - 3$ is ☐.

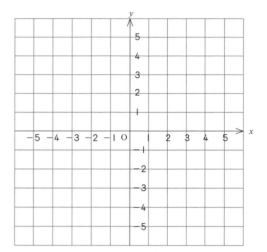

If you need to, make a table with the values of x and y.

3 **Determine the slope of each equation.**

8 points per question

Example

The slope of $y = 4x - 6$ is 4.

(1) The slope of $y = 2x + 8$ is ☐.

(4) The slope of $y = \dfrac{5}{2}x$ is ☐.

(2) The slope of $y = 3x - \dfrac{1}{2}$ is ☐.

(5) The slope of $y = \dfrac{3}{8}x - \dfrac{7}{6}$ is ☐.

(3) The slope of $y = x + 6$ is ☐.

Do your best!

Graphs
Slope

Date   /   /

Name

Level ☆☆

Score
/ 100

■ The Answer Key is on page 189.

1 **Graph each equation to find the slope.**

7 points per question

Examples

Lines（1）and（2）have a **negative slope**.

The slope of line（1）is -3 because the line falls 3 units vertically for every 1 unit it runs horizontally.

The slope of line（2）is $-\frac{1}{2}$ because the line falls 1 unit vertically for every 2 units it runs horizontally.

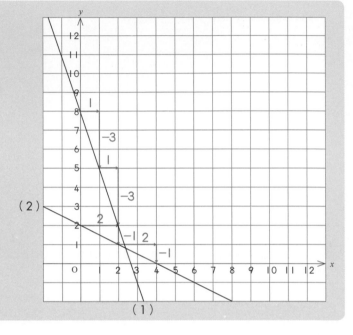

（1） The slope of $y = -2x + 2$ is ☐ .

x	0	1
y		

（2） The slope of $y = -x + 3$ is ☐ .

x	0	1
y		

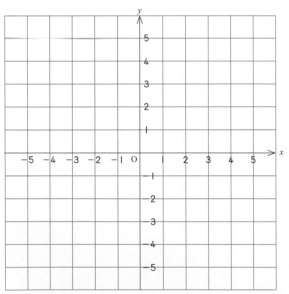

（3） The slope of $y = -\frac{1}{2}x - 1$ is ☐ .

x	0	1	2
y			

150   © Kumon Publishing Co., Ltd.

2   Graph each equation to find the y-intercept.

8 points per question

(1)   The y-intercept of $y=-3x$ is ☐.

(2)   The y-intercept of $y=-x+\dfrac{1}{2}$ is ☐.

(3)   The y-intercept of $y=\dfrac{1}{2}x-2$ is ☐.

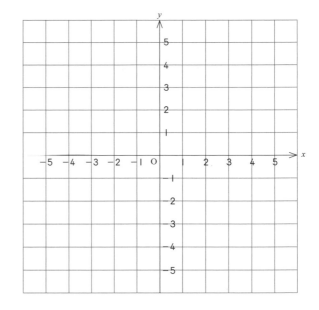

3   Determine the slope and y-intercept of each equation.

11 points per question

(1) $y=-3x+1$ slope: -3 y-intercept: _____

(2) $y=-2x-6$ slope: _____ y-intercept: _____

(3) $y=-\dfrac{1}{2}x-2$ slope: _____ y-intercept: _____

(4) $y=-\dfrac{5}{2}x$ slope: _____ y-intercept: _____

(5) $y=-6x-\dfrac{3}{5}$ slope: _____ y-intercept: _____

You understand slope and y-intercept very well!

Graphs

Date   /   /

Name

Level
★★

Score
/100

■ The Answer Key is on page 189.

1 **Graph each equation to find the slope and y-intercept.**

10 points per question

(1) $y = 3x - 1$

slope:_____ y-intercept:_____

(3) $y = -2x + 3$

slope:_____ y-intercept:_____

(2) $y = x + 2$

slope:_____ y-intercept:_____

(4) $y = -x - 4$

slope:_____ y-intercept:_____

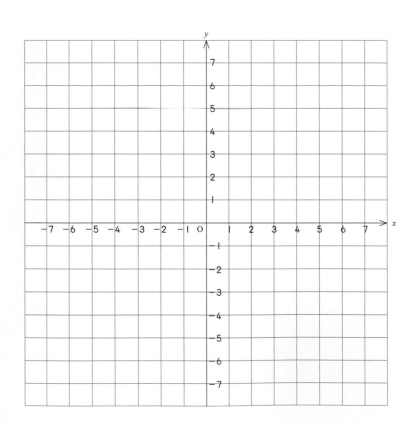

2 **Graph each equation to find the slope and y-intercept.** 15 points per question

(1) $y = \dfrac{3}{2} x + 1$

slope: _____ y-intercept: _____

(2) $y = \dfrac{1}{2} x - \dfrac{3}{2}$

slope: _____ y-intercept: _____

(3) $y = -\dfrac{2}{3} x - 2$

slope: _____ y-intercept: _____

(4) $y = -\dfrac{5}{2} x + \dfrac{1}{2}$

slope: _____ y-intercept: _____

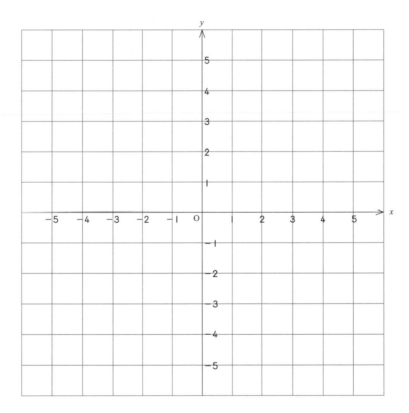

Practice makes perfect!

■ The Answer Key is on page 189.

1 **Use the graph to complete each statement.**

5 points per question

(1) When the value of x is 0, the value of y is ☐.

(2) When the value of x is ☐, the value of y is 0.

(3) Therefore, when x changes from -3 to 0,

  y changes from 0 to ☐.

(4) Thus, as the value of x increases by 3,

  the value of y increases by ☐.

(5) In other words, as the value of x increases by 1,

  the value of y increases by ☐.

(6) The slope of the line is ☐.

(7) The y-intercept is ☐.

(8) The equation of the line is $y = $ ☐ $x + $ ☐.

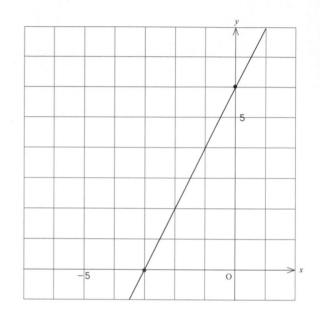

Don't forget!

If the relation between x and y can be expressed as $y = mx + b$, then the equation is called a **linear function**.

2. Use the graph to complete each statement.

5 points per question

(1) When the value of x is 0, the value of y is ☐.

(2) When the value of x is ☐, the value of y is 0.

(3) Therefore, when x changes from 0 to 6,

   y changes from ☐ to 0.

(4) Thus, as the value of x increases by 6,

   the value of y changes by ☐.

   In other words, as the value of x increases

   by 1, the value of y changes by ☐.

(5) The slope of the line is ☐, and the y-intercept is ☐.

(6) The equation of the line is $y =$ ☐ $x +$ ☐.

3. Use the graph to complete each statement.

5 points per question

(1) When the value of x is 0, the value of y is ☐.

(2) When the value of x is ☐, the value of y is 0.

(3) Therefore, when x changes from -2 to 0,

   y changes from 0 to ☐.

(4) Thus, as the value of x increases by 2,

   the value of y changes by ☐.

   In other words, as the value of x increases by 1,

   the value of y changes by ☐.

(5) The slope of the line is ☐. The y-intercept is ☐.

(6) The equation of the line is: $y =$ ☐ $x -$ ☐.

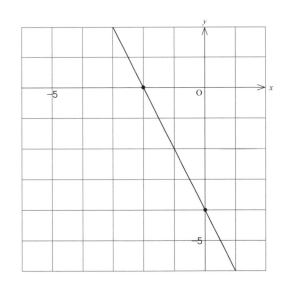

You're off to a great start with linear functions!

Linear Functions

30

Date: / /

Name:

Level ☆☆

Score: /100

■ The Answer Key is on page 189.

 1   **Use the slope and y-intercept to graph each line.**

10 points per question

(1)   slope: 2, y-intercept: -3

(2)   slope: 1, y-intercept: 0

(3)   slope: -3, y-intercept: -1

(4)   slope: $-\dfrac{3}{2}$, y-intercept: -3

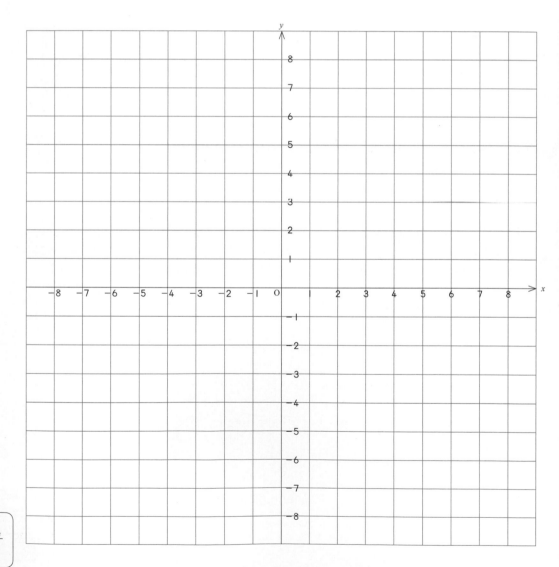

slope $= \dfrac{\text{rise}}{\text{run}}$

2 Use the slope and y-intercept to graph each line.

(1) slope: 3, y-intercept: 3

(2) slope: -4, y-intercept: 0

(3) slope: $-\dfrac{5}{2}$, y-intercept: $-\dfrac{3}{2}$

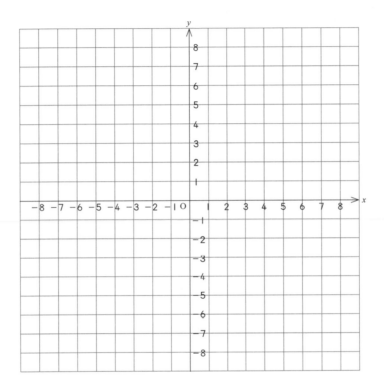

3 Determine the equation of each line above.

(1) $y = \boxed{}\,x + \boxed{}$

(2) $y = \text{—}$

(3) $y =$

You're doing so well!

Date   /   /

Name

■ The Answer Key is on page 190.

1   **Determine the equation of the line by completing each exercise.**   4 points per question

(1)   The slope of the line is ☐.

(2)   The y-intercept of the line is ☐.

(3)   Therefore, the equation of the line is:

$$y = \boxed{}\, x - \boxed{}.$$

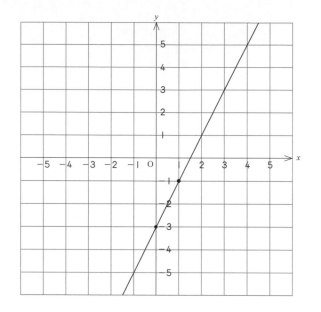

2   **Determine the equation of each line on the graph.**   8 points per question

(1) $y =$

(2) $y =$

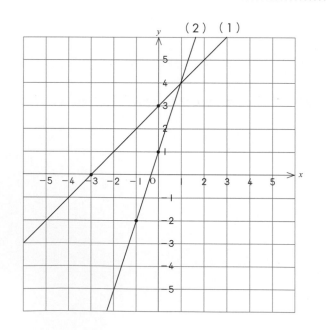

3 **Determine the equation of each line on the graph below.** 12 points per question

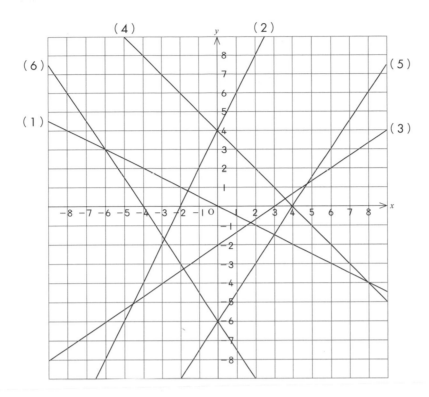

(1) $y =$

(2) $y =$

(3) $y =$

(4) $y =$

(5) $y =$

(6) $y =$

Good going!

32
Linear Functions

Date   /   /

Name

Level
★★

Score

/100

■ The Answer Key is on page 190.

1   **Graph each line and determine the equation.**

25 points per question

(1)   A line that passes through (1, 5) and has a slope of 2.

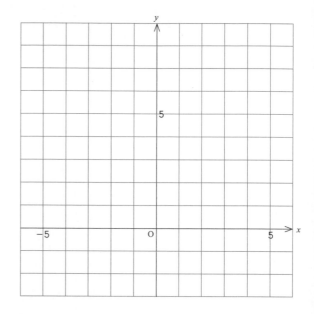

The y-intercept is ☐.

The equation of the line is: $y = 2x + $ ☐.

(2)   A line that passes through (−4, 3) and has a slope of $-\dfrac{1}{2}$.

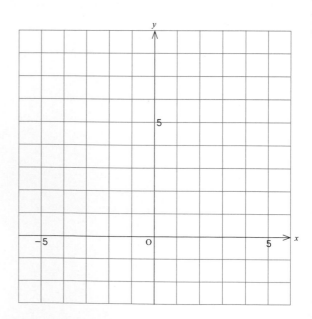

The y-intercept is ☐.

The equation of the line is:

2 **Determine the equation of each line and then graph it.**

(1) A line that passes through (1, 4) and has a slope of -3.

Because the slope is -3,

$y = -3x + b \quad \cdots ①$

Because the line passes through (1, 4),

substitute (1, 4) into ①:

$4 = -3 \times \boxed{} + b$

$b = \boxed{}$

Substitute this into ①,

$y = -3x + \boxed{}$

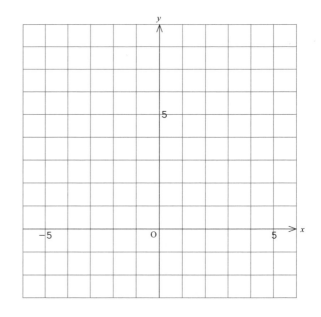

(2) A line that passes through $(-1, -2)$ and has a slope of $\dfrac{3}{2}$.

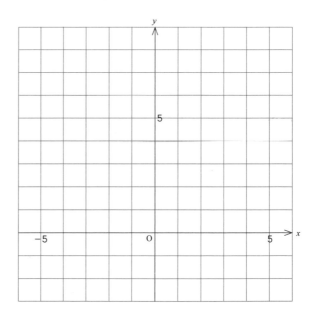

 The letter b is used to represent an unknown value of the y-intercept.

 Just superb!

161

Linear Functions

Date / /

Name

■ The Answer Key is on page 190.

1 **Find the slope of each line that passes through the given points.**   10 points per question

(1)

(1, 2) , (4, 7)

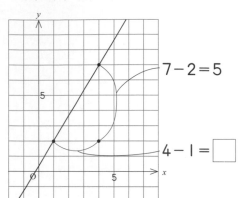

$7 - 2 = 5$

$4 - 1 = \boxed{}$

slope $= \dfrac{5}{\boxed{}}$

(3)

(2, 5) , (4, −3)

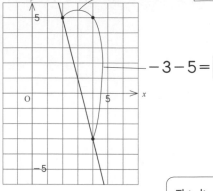

$4 - 2 = \boxed{}$

$-3 - 5 = \boxed{}$

slope $= -\dfrac{\boxed{}}{}$

This line has a negative slope.

(2)

(−3, −2) , (1, 4)

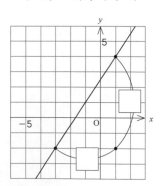

slope =

(4)

(−3, 2) , (1, −6)

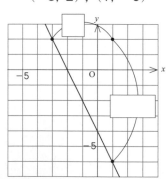

slope =

2 **Find the slope of the line that passes through the given points.**   10 points per question

Don't forget!

The slope of a line that passes through the points $(a,\ b)$ and $(c,\ d)$ is:

$$m = \frac{d-b}{c-a}$$

The letter m is used to represent slope.

Examples

$(1,\ 2),\ (4,\ 7)$

$$m = \frac{7-2}{4-1} = \frac{5}{3}$$

$(5,\ -1),\ (7,\ -3)$

$$m = \frac{-3-(-1)}{7-5} = -\frac{2}{2} = -1$$

(1) $(2,\ 8),\ (6,\ 9)$

$$m = \frac{9-8}{6-\boxed{}} = \frac{1}{\boxed{}}$$

(4) $(7,\ 0),\ (10,\ -2)$

$$m =$$

(2) $(0,\ -2),\ (4,\ 5)$

$$m =$$

(5) $(1,\ 1),\ (-1,\ 7)$

$$m =$$

(3) $(2,\ -6),\ (8,\ -3)$

$$m =$$

(6) $(-9,\ -5),\ (-3,\ -9)$

$$m =$$

Reduce the fraction if possible.

You're flying now!

Level ☆☆

Date / /

Name

Score /100

■ The Answer Key is on page 190.

1 **A line passes through the points** $(0, -1)$ **and** $(3, 5)$**. Use this information to answer each exercise.**

25 points per question

(1) Determine the equation of the line.

Calculate the slope:

$$m = \frac{5 - (\boxed{})}{3 - \boxed{}} =$$

Let the equation of the line be:

$$y = \boxed{}\,x + b \quad \cdots \text{①}$$

Substitute $(0, -1)$ into ①:

$$\boxed{} = \boxed{} \times 0 + b$$

$$b = \boxed{}$$

Substitute this into ①:

$$y =$$

You could also substitute $(3, 5)$ into ①.

(2) Graph the equation.

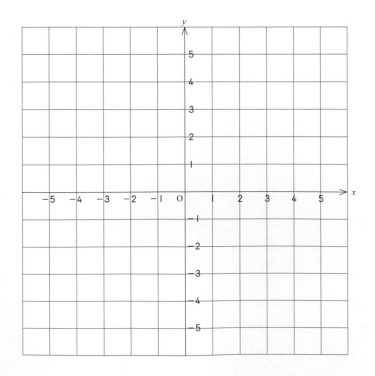

© Kumon Publishing Co., Ltd.

(2) Find the equation of each line that passes through the given points, and then graph the equation.

25 points per question

(1) $(-1, 5)$, $(-2, -3)$

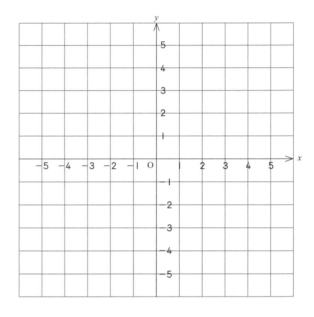

(2) $(-2, 3)$, $(4, -1)$

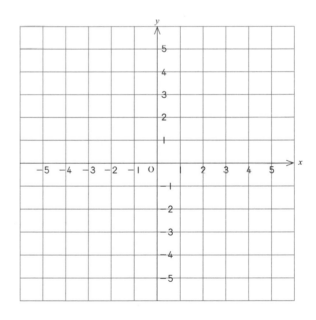

Carefully follow through on each step.

Linear Functions

35

Date   /   /

Name

Level
☆☆

Score

/100

■ The Answer Key is on page 190.

1 **Find the equation of each line.**

12 points per question

(1) A line that passes through $(2, 5)$ and has a slope of -1.

〈**Ans.**〉 _____

(2) A line that passes through $(-6, -1)$ and has a slope of $\dfrac{1}{2}$.

〈**Ans.**〉 _____

(3) A line that passes through the origin and has a slope of $-\dfrac{5}{3}$.

〈**Ans.**〉 _____

The origin is where the x-axis and y-axis meet, therefore the coordinates of the origin are: $(0, 0)$.

2 **Find the equation of each line.**

16 points per question

(1) A line that passes through (1, 6) and (3, 4).

⟨**Ans.**⟩ _____

(2) A line that passes through (−4, −3) and (2, 0).

⟨**Ans.**⟩ _____

(3) A line that passes through (−1, 5) and the origin.

⟨**Ans.**⟩ _____

(4) A line that passes through (1, 8) and has a y-intercept of 2.

Because the line has a y-intercept of 2, the line passes through ($\boxed{}$, 2).

⟨**Ans.**⟩ _____

You've got this down, so let's move on to something new!

36

Linear Functions

Level

Score
/ 100

■ The Answer Key is on page 191.

1 **Graph each equation by finding the slope and y-intercept.**

20 points per question

(1) $2x + y = 5$

Transpose $2x$ to the right of the equal sign.

$y = \boxed{} + 5$

Therefore, the slope is $\boxed{}$,

and the y-intercept is $\boxed{}$.

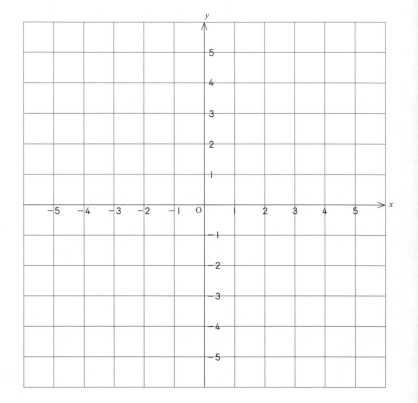

(2) $-3x + y = -4$

The slope is $\boxed{}$.

The y-intercept of the line is $\boxed{}$.

(3) $\dfrac{1}{2}x + y = -1$

The slope is $\boxed{}$.

The y-intercept of the line is $\boxed{}$.

Remember to label each line.

2 **Graph each equation by finding the slope and y-intercept.**

20 points per question

(1) $2x + 3y = 12$

Transpose $2x$ to the right of the equal sign.

$$3y = \boxed{} + 12$$

$$y = \boxed{} + 4$$

Therefore, the slope is $\boxed{}$,

and the y-intercept is $\boxed{}$.

(2) $-6x - 2y = 1$

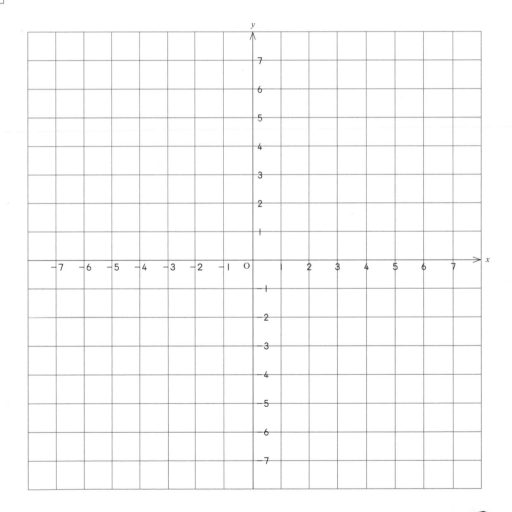

You are doing stellar work!

37

Simultaneous Linear Equations

Date   /   /

Name

■ The Answer Key is on page 191.

1 **Use the simultaneous linear equations** $\begin{cases} 3x+y=-1 & \cdots① \\ -x+y=3 & \cdots② \end{cases}$ **for each exercise.**

25 points per question

(1) Find the point of intersection by graphing.

Rewrite ①: $y = \boxed{} -1$ ···③

Rewrite ②: $y = $ ···④

Graph ③ and ④:

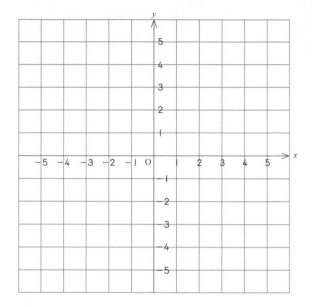

Point of intersection: $(x, y) = ($, $)$

┌─ **Don't forget!** ─────────────────────────────┐
│ The **point of intersection** is a point where two or more lines cross. │
└──┘

(2) Find the point of intersection by solving the simultaneous linear equations algebraically.

$\begin{cases} 3x+y=-1 & \cdots① \\ -x+y=3 & \cdots② \end{cases}$

Point of intersection: $(x, y) = ($, $)$

You can find the point of intersection by graphing both equations or solving them algebraically.

2 **Use the simultaneous linear equations** $\begin{cases} x - 2y = 7 & \cdots ① \\ -3x + y = -6 & \cdots ② \end{cases}$ **for each exercise.**

25 points per question

（1） Find the point of intersection by graphing.

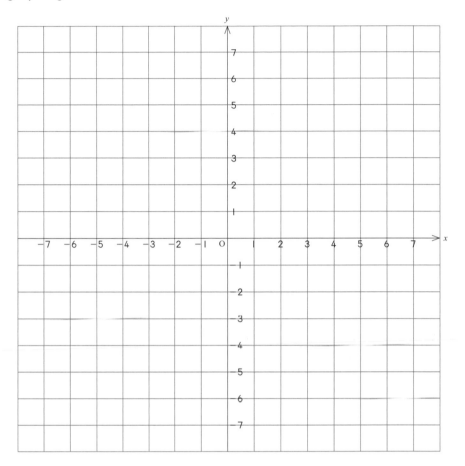

Point of intersection: $(x, y) = ($, $)$

（2） Find the point of intersection by solving the simultaneous linear equations algebraically.

Point of intersection: $(x, y) = ($, $)$

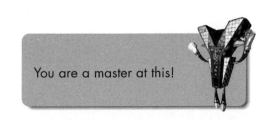

You are a master at this!

Level ☆☆☆

■ The Answer Key is on page 191.

Don't forget!

The equation for a **horizontal line** is $y = a$ (where a represents a number). Horizontal lines have a slope of 0.

Example

$y = 3$

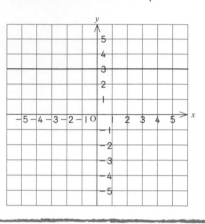

---

1   **Graph each equation.**

10 points per question

(1) $y = -3$

(2) $y - 4 = 0$

Rewrite the equation: $y = \boxed{}$

(3) $2y - 3 = 0$

(4) $2y + 1 = 0$

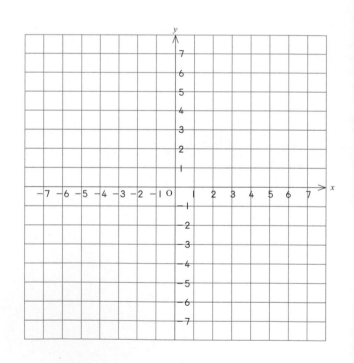

Don't forget!

The equation for a **vertical line** is $x = a$ (where a represents a number). The slope of a vertical line does not exist.

Example

$$x = -2$$

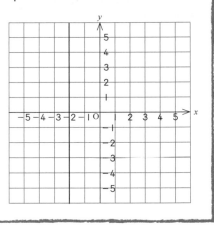

---

2 **Graph each equation.**

10 points per question

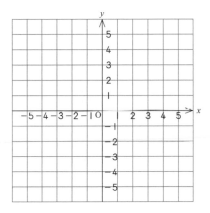

(1) $x = 4$

(2) $x + 5 = 0$

(3) $3x + 6 = 0$

3 **Graph each equation.**

10 points per question

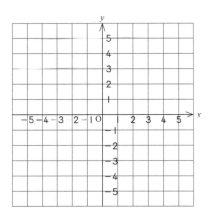

(1) $2x + y = 3$

(2) $2x = 3$

(3) $y = 3$

Keep up the good work!

Parallel Lines

39

Date / /

Name

Score /100

■ The Answer Key is on page 192.

1 **Graph the lines** $\begin{cases} y = 2x + 4 & \cdots ① \\ y = 2x - \dfrac{1}{2} & \cdots ② \end{cases}$ **and answer the question.** 25 points for completion

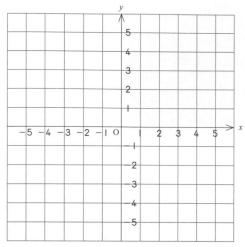

How many points of intersection do the two lines have? The two lines have ☐ points of intersection.

Don't forget!

Parallel lines are lines that have the same slope but different y-intercepts. Parallel lines do not have any points of intersection.

---

2 **Graph** $\begin{cases} 6x + 3y = 12 & \cdots ① \\ -4x - 2y = 1 & \cdots ② \end{cases}$ **to show that they are parallel lines.** 25 points for completion

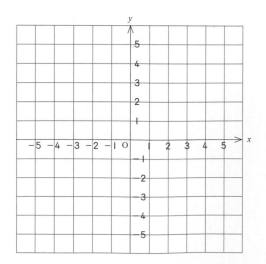

3 **Find the equation of the line that is parallel to the given line and passes through the given point.**

25 points per question

(1) $y = 2x + 3$; $(-1, -6)$

Because the line is parallel to $y = 2x + 3$,

the slope of the line is also $\boxed{}$.

Because the slope is $\boxed{}$,

$y = \boxed{} x + b$.

Because the line passes through $(-1, -6)$,

⟨**Ans.**⟩ _____

(2) $x + 4y = -6$; $(8, -3)$

⟨**Ans.**⟩ _____

You can figure out any line!

40

Perpendicular Lines

Date / /

Name

■ The Answer Key is on page 192.

1 **Graph each equation.**

10 points per question

(1) $y = \dfrac{2}{3}x + 1$

(2) $y = -\dfrac{3}{2}x - 2$

(3) $y = -\dfrac{3}{2}x + 7$

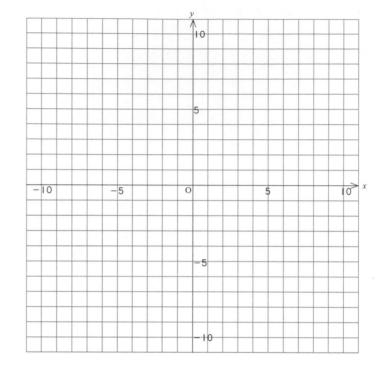

 The slope of line (1) has the opposite sign and is the reciprocal of the slope of line (2); therefore line (1) is perpendicular to line (2).

2 **Given that line ① is perpendicular to line ②, complete each exercise.**

10 points per question

(1) If the slope of line ① is $\dfrac{3}{4}$, then the slope of line ② is $\boxed{-\dfrac{4}{3}}$.

(2) If the slope of line ① is $-\dfrac{7}{2}$, then the slope of line ② is $\boxed{}$.

(3) If the slope of line ① is 6, then the slope of line ② is $\boxed{}$.

3 Find the equation of the line that is perpendicular to each given line and passes through each given point.

20 points per question

(1) $y = \dfrac{2}{3}x + 5$; $(-4, -1)$

Because the line is perpendicular to $y = \dfrac{2}{3}x + 5$, the slope of the line is ☐.

Because the slope is ☐, $y = ☐x + b$.

Because the line passes through $(-4, -1)$,

⟨Ans.⟩ _____

(2) $4x + 8y = 3$; $(-6, -7)$

Rewrite the equation:

$8y = -4x + 3$

$y =$

⟨Ans.⟩ _____

I'm impressed!

© Kumon Publishing Co., Ltd.

Perpendicular Lines

Level

Score

/100

Date / /

Name

■ The Answer Key is on page 192.

1 **Find the equation of the line that is perpendicular to the given line and passes through the given point.**

25 points per question

(1) $3x - 2y = 5$; $(-3, -1)$

〈Ans.〉 _____

(2) $-5x = \dfrac{1}{2}y - 1$; $\left(\dfrac{1}{4},\ \dfrac{1}{2}\right)$

〈Ans.〉 _____

2 The equation for Line A is $-6x + 4y = -8$. **Use this information to find the equation for each line, and then graph each line below.**

25 points per question

(1) Line B is parallel to Line A and passes through $(-3, 1)$.

⟨**Ans.**⟩ _____

(2) Line C is perpendicular to Line A and passes through $(-3, 1)$.

⟨**Ans.**⟩ _____

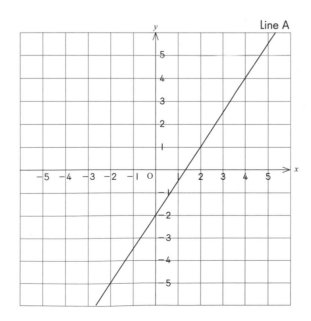

Fantastic work!

Review

Date / /

Name

Level
☆☆

Score

/100

■ The Answer Key is on page 192.

1 Use the following method to solve the equations $\begin{cases} 6x + 5y = 8 & \cdots ① \\ -4x + 3y = 20 & \cdots ② \end{cases}$. 10 points per question

(1) Addition/subtraction method

(2) Substitution method

⟨Ans.⟩ _____

⟨Ans.⟩ _____

2 Answer the word problem.

14 points for completion

Nori's farm has several rabbits and chicks. There are 3 times as many chicks as rabbits. If each member of Nori's family holds 2 rabbits, there are 3 rabbits left over. If each family member holds 3 chicks, there are 27 chicks left over. How many people make up Nori's family? How many chicks are there?

⟨Ans.⟩   people in Nori's family

chicks

3 Solve each inequality.

10 points per question

(1) $2(x+5)-\dfrac{1}{2}x<2$

(3) $-\dfrac{9-x}{2}<\dfrac{-3-x}{4}$

(2) $-3x-2\left(\dfrac{1}{4}x-\dfrac{1}{2}\right)\geq 2-x$

(4) $\dfrac{3x+2}{5}\geq\dfrac{-x+4}{2}$

4 Solve each inequality or find the range of x that satisfies both inequalities.

13 points per question

(1) $\begin{cases} 3x>-x+8 \\ 5-x>x+9 \end{cases}$

(2) $\begin{cases} 4\left(3-\dfrac{1}{2}x\right)\geq 3(1+2x) \\ -\left(-\dfrac{1}{2}x-1\right)\leq 2(1-3x) \end{cases}$

⟨Ans.⟩ _____

⟨Ans.⟩ _____

You're almost at the finish line!

Review

Date   /   /

Name

Level

Score
/100

■ The Answer Key is on page 192.

1 **Graph each equation.**

15 points per question

(1) $3x + y = 2$

(2) $4x - y = 0$

(3) $3x + 5y = 0$

(4) $-2x + 6y = 3$

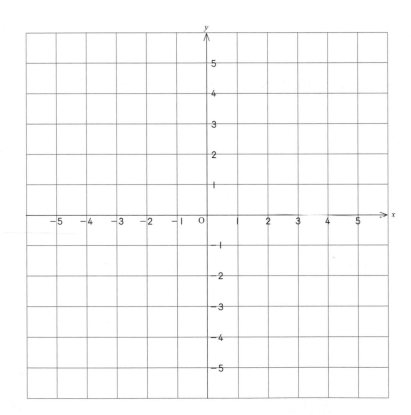

2 Use each method to find the point of intersection of

20 points per question

$$\begin{cases} 2x - 4y = -14 & \cdots① \\ 12x + 8y = 12 & \cdots② \end{cases}$$

(1) Graph the equations to find the point of intersection.

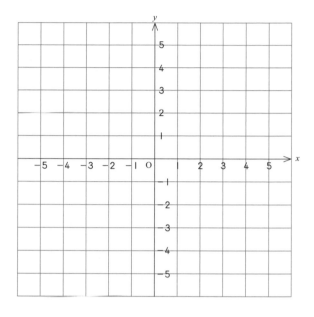

Point of intersection: $(x, y) = ($, $)$

(2) Use algebra to find the point of intersection.

Point of intersection: $(x, y) = ($, $)$

Congratulations! You completed
Algebra Workbook II!

1 **Solving Equations Review** pp 98, 99

1 (1) $x = 11$

(2) $x = 6$

(3) $x = 1$

(4) $x = -8$

(5) $x = -4$

(6) $x = -\dfrac{3}{4}$

(7) $x = \dfrac{7}{3}$

(8) $x = \dfrac{1}{12}$

(9) $x = -\dfrac{11}{3}$

(10) $x = \dfrac{11}{6}$

2 (1) $x = 5$

(2) $x = -24$

(3) $x = -\dfrac{4}{3}$

(4) $x = -1$

(5) $x = 10$

(6) $x = \dfrac{15}{2}$

(7) $x = -\dfrac{9}{2}$

(8) $x = -\dfrac{1}{15}$

(9) $x = \dfrac{2}{3}$

(10) $x = \dfrac{1}{2}$

2 **Solving Equations Review** pp 100, 101

1 (1) $x = 3$

(2) $x = 2$

(3) $x = -4$

(4) $x = -5$

(5) $x = \dfrac{3}{2}$

(6) $x = -\dfrac{14}{5}$

(7) $x = -\dfrac{7}{2}$

(8) $x = -\dfrac{15}{4}$

2 (1) $x = 7$

Left side $= 3 \times 7 - 6 = 15$

Right side $= 7 + 8 = 15$

(2) $x = 3$

Left side $= 2 \times 3 + 15 = 21$

Right side $= 9 \times 3 - 6 = 21$

(3) $x = 1$

Left side $= -\dfrac{1}{3} \times 1 + \dfrac{1}{2} = \dfrac{1}{6}$

Right side $= \dfrac{5}{3} \times 1 - \dfrac{3}{2} = \dfrac{1}{6}$

(4) $x = 4$

Left side $= -\dfrac{1}{2} \times 4 - \dfrac{1}{3} = -\dfrac{7}{3}$

Right side $= -\dfrac{3}{4} \times 4 + \dfrac{2}{3} = -\dfrac{7}{3}$

3 **Solving Equations Review** pp 102, 103

1 (1) $x = -5$

(2) $x = 2$

(3) $x = \dfrac{9}{5}$

(4) $x = -\dfrac{2}{3}$

(5) $x = -\dfrac{21}{2}$

2 (1) $x = y - 3$

(2) $x = -b - c$

(3) $x = -w + y$

(4) $x = -f + 3$

(5) $x = -w - 8$

(6) $x = -\dfrac{b}{4}$

(7) $x = \dfrac{b}{a}$

(8) $x = \dfrac{4a + b}{9}$

(9) $x = \dfrac{-2b + 3c}{a}$

(10) $x = \dfrac{s + 4t}{r}$

4 **Simultaneous Linear Equations Review** pp 104, 105

1 (1) Ans. $(x, y) = (-1, 2)$

(2) Ans. $(x, y) = (-4, -3)$

(3) Ans. $(x, y) = \left(\dfrac{1}{2}, -4\right)$

(4) Ans. $(x, y) = \left(-3, -\dfrac{2}{3}\right)$

2 (1) Ans. $(x, y) = (-1, -2)$

(2) Ans. $(x, y) = \left(-\dfrac{3}{5}, 3\right)$

(3) Ans. $(x, y) = \left(4, -\dfrac{2}{3}\right)$

(4) Ans. $(x, y) = (-6, -2)$

5 **Simultaneous Linear Equations Review** pp 106, 107

1 (1) Ans. $(x, y) = (2, -3)$

(2) Ans. $(x, y) = (2, 1)$

2 (1) Ans. $(x, y) = (-4, 2)$

① : $\boxed{-4} - 2 \times \boxed{2} = -8$

② : $2 \times \boxed{(-4)} + 3 \times \boxed{2} = -2$

(2) Ans. $(x, y) = \left(-\dfrac{1}{2}, \dfrac{2}{3}\right)$

6 **Simultaneous Linear Equations Review** pp 108, 109

1 (1) Ans. $(x, y) = (2, -2)$

(2) $\boxed{4}$ Ans. $(x, y) = (3, -1)$

2 (1) Ans. $(x, y) = (4, 2)$

(2) Ans. $(x, y) = (4, 3)$

(3) Ans. $(x, y) = (5, -4)$

7 Simultaneous Linear Equations Review pp 110, 111

1 (1) **Ans.** $(x, y)=(-2, -5)$ (3) **Ans.** $(x, y)=(4, -3)$

(2) **Ans.** $(x, y)=(-7, -2)$ (4) **Ans.** $(x, y)=(1, 1)$

2 (1) $\boxed{-2}$ (3) $\boxed{-4x-1}$

 Ans. $(x, y)=(-3, 4)$ **Ans.** $(x, y)=(2, -3)$

(2) **Ans.** $(x, y)=(5, 1)$ (4) **Ans.** $(x, y)=(3, 1)$

8 Simultaneous Linear Equations Review pp 112, 113

1 (1) **Ans.** $(x, y)=(1, -3)$ (2) **Ans.** $(x, y)=(1, -3)$

2 (1) **Ans.** $(x, y)=(-2, -5)$ (2) **Ans.** $(x, y)=(-2, -5)$

3 (1) **Ans.** $(x, y)=(-5, -3)$ (2) **Ans.** $(x, y)=(-5, -3)$

4 (1) **Ans.** $(x, y)=(4, -6)$ (2) **Ans.** $(x, y)=(4, -6)$

9 Simultaneous Linear Equations in Three Variables pp 114, 115

1 (1) $\boxed{5}$, $\boxed{2}$, $\boxed{-2}$, $\boxed{3}$, $\boxed{3}$, $\boxed{5}$, $\boxed{-4}$, $\boxed{2}$, $\boxed{3}$, $\boxed{2}$, $\boxed{3}$,
 $\boxed{2}$, $\boxed{1}$ **Ans.** $(x, y, z)=(\boxed{1}, \boxed{3}, \boxed{2})$

(2) **Ans.** $(x, y, z)=(-3, -2, 2)$

2 (1) $\boxed{11}$, $\boxed{9}$, $\boxed{-5}$ **Ans.** $(x, y, z)=(4, -2, 1)$

(2) **Ans.** $(x, y, z)=(1, -4, 1)$

10 Simultaneous Linear Equations in Three Variables pp 116, 117

1 (1) **Ans.** $(x, y, z)=(3, 1, 2)$

(2) $\boxed{y}$ **Ans.** $(x, y, z)=(-5, -1, 4)$

2 (1) $\boxed{3}$, $\boxed{6}$ **Ans.** $(x, y, z)=(-2, 1, -2)$

(2) $\boxed{y}$ **Ans.** $(x, y, z)=(6, 2, 3)$

11 Simultaneous Linear Equations in Three Variables pp 118, 119

1 $\boxed{15}$, $\boxed{5}$, $\boxed{25}$, $\boxed{25}$, $\boxed{-45}$, $\boxed{-16}$, $\boxed{-24}$, $\boxed{-4}$,
 $\boxed{-6}$ **Ans.** $(x, y, z)=(-3, 2, 2)$

2 **Ans.** $(x, y, z)=(-2, -4, 1)$

12 Word Problems with Simultaneous Linear Equations pp 120, 121

1 (1) $\begin{cases} 6x+7=y \\ \boxed{8}x+1=y \end{cases}$

 Ans. $\boxed{3}$ benches
  $\boxed{25}$ students

(2) Let x be the number of train cars
 and y be the number of students:
 $\begin{cases} 3x+16=y \\ 5x=y \end{cases}$

 Ans. 8 train cars
  40 students

2 (1) $\boxed{2}$, $\boxed{3}$, $\boxed{3}$

 Ans. 4 students
  9 apples

(2) Let x be the number of students
 and y be the number of pens:
 $\begin{cases} 2x+6=y \\ 5x+7=2y \end{cases}$

 Ans. 5 students
  16 pens

13 Word Problems with Simultaneous Linear Equations pp 122, 123

1 (1) $5x=\boxed{20}$ **Ans.** 4 hours

(2) $\boxed{2}$, $\boxed{8}$, $\boxed{8}x=40$ **Ans.** 5 hours

(3) $15x=45$ **Ans.** 3 hours

2 $(x-\boxed{y})\times 5=40$
  $(x+\boxed{y})\times 2=60$

 Ans. Jane's speed: 19 miles per hour
  The current's speed: 11 miles per hour

14 Inequalities pp 124, 125

1 (1) $x>6$ (4) $x>5$

(2) $x>-3$ (5) $x>4$

(3) $x>4$ (6) $x>\dfrac{7}{2}$

2 (1) $2x>12$ (3) $x>1$
  $x>6$

(2) $x>11$ (4) $x>0$

3 (1) $x < 10$

(2) $x < 13$

(3) $x < -2$

(4) $x < 9$

(5) $x < \dfrac{13}{5}$

(6) $2x < -3$

$x < -\dfrac{3}{2}$

(7) $x < -4$

(8) $x < \dfrac{7}{2}$

(9) $x < -\dfrac{1}{9}$

4 (1) $x \boxed{>} -5$

(2) $x \boxed{<} 3$

(3) $x > -2$

(4) $x > 2$

(5) $\boxed{-8}$, $x < 4$

(6) $x < -2$

(7) $x > 14$

(15) Inequalities
pp 126, 127

1 (1) $x \geq 1$

(2) $x \geq \dfrac{5}{3}$

(3) $\boxed{2}$, $\boxed{6}$, $x \leq 4$

(4) $x \leq 0$

(5) $x \leq 3$

(6) $x \geq -2$

(7) $x \geq \dfrac{3}{4}$

(8) $x \leq -\dfrac{8}{23}$

2 (1) $x < -26$

(2) $x \leq -\dfrac{16}{3}$

(3) $x \leq \dfrac{23}{13}$

(4) $x \geq 5$

(5) $x \leq 32$

(6) $x > -\dfrac{26}{3}$

(16) Word Problems with Inequalities
pp 128, 129

1 (1) $x + 6 > \boxed{4}x$ 　　　Ans. $x < 2$

(2) $y - 8 \leq 6$ 　　　Ans. $y \leq 14$

(3) $5z \geq z + 6$ 　　　Ans. $z \geq \dfrac{3}{2}$

(4) $-2w \leq w - 3$ 　　　Ans. $w \geq 1$

(5) $3 + q > 8q$ 　　　Ans. $q < \dfrac{3}{7}$

2 (1) $\boxed{3}x + 2 < \boxed{10}$ 　　　Ans. $x < \dfrac{8}{3}$

(2) $\boxed{3}(x + \boxed{2}) < 10$ 　　　Ans. $x < \dfrac{4}{3}$

(3) $2(3x + 5) \geq 9x$ 　　　Ans. $x \leq \dfrac{10}{3}$

(4) $\dfrac{1}{2}(x - \boxed{4}) < 2(\boxed{5} - x)$ 　　　Ans. $x < \dfrac{24}{5}$

(17) Inequalities Range of x
pp 130, 131

1 (1)

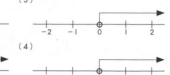

(2)

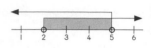

(3)

(4)

2 (1)

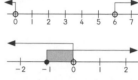

(2)

3 (1) Ans. $\boxed{2} < x < \boxed{5}$

(2) Ans. No solution

(3) Ans. $\boxed{-1} \leq x < \boxed{0}$

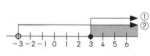

(4) Ans. $\boxed{-10} \leq x \leq \boxed{-7}$

(18) Inequalities Range of x
pp 132, 133

1 (1) Ans. $x < -1$

(2) Ans. $x \geq 3$

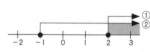

(3) Ans. $x < 0$

(4) Ans. $x \geq 2$

(5) Ans. $x \leq -4$

2 (1) Ans. $2 < x < 8$

(2) Ans. No solution

(3) Ans. $x \geq 2$

(4) Ans. No solution

(5) Ans. $x < -6$

(6) $\boxed{8}$, $\boxed{3}$, Ans. $\boxed{3} \leq x < \boxed{8}$

(7) Ans. $x < -6$

(8) Ans. No solution

⑲ Word Problems with Inequalities pp 134, 135

1 (1) $\boxed{x}$, $50 - x \leq \boxed{4}(20 - \boxed{x})$

$x \leq 10$ **Ans.** 10 coins or fewer

(2) Let x be the number of toys Maria and Alex each

gave Jane.

$40 - x \geq \dfrac{1}{2}(60 - x)$

$x \leq 20$ **Ans.** 20 toys or fewer

2 (1) $\boxed{2}$, $x + \boxed{2}x > 60$ **Ans.** more than 20 boys

(2) Let x be the amount of radios.

$x + 2x \leq 285$

$x \leq 95$

Ans. less than or equal to 95 radios

⑳ Graphs pp 136, 137

1 (1) $\boxed{1}$

(2) $\boxed{-2}$

(3) $\boxed{-1}$

(4) $\boxed{0}$

(5) $\boxed{-4}$

(6) $\boxed{-3}$, $\boxed{-1}$

(7) $\boxed{-3}$

2 (1) $\boxed{2}$

(2) $\boxed{4}$

(3) $\boxed{\dfrac{1}{2}}$

(4) $\boxed{-\dfrac{5}{2}}$, $\boxed{-\dfrac{3}{2}}$

(5) $\boxed{0}$

3 (1) $(-4, -2)$

(2) $(2, 3)$

(3) $(0, 0)$

(4) $\left(\dfrac{5}{2}, -3\right)$

(5) $\left(-\dfrac{7}{2}, \dfrac{5}{2}\right)$

(6) $\boxed{C}$

㉑ Graphs pp 138, 139

1
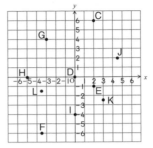

2
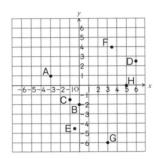

(9) $\boxed{H}$

(10) $\boxed{B}$

㉒ Graphs pp 140, 141

1 (1) $2 \times (-3) - 3 = -9$

(2) -7

(3) -5

(4) -3

(5) -1

(6) 1

(7) 3

(8) -2

(9) $-\dfrac{3}{2}$

(10) -6

2 (1) ① -4

② -2

③ -1

④ $-\dfrac{5}{2}$

(2)
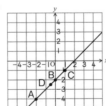

3

x	-1	$-\dfrac{1}{2}$	0	$\dfrac{1}{2}$	1	2
y	3	2	1	0	-1	-3

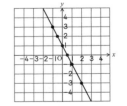

23 Word Problems with Graphs pp 142, 143

1 (1)

x (hour)	0	1	2	2.5	3	4
y (mile)	0	3	6	7.5	9	12

(2) $y = \boxed{3}x$

(3)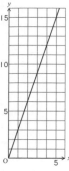

(4) **Ans.** 15 miles

2 (1)

x (hour)	1	2	5	8	10
y (mile)	12	24	60	96	120

(2) $y = 12x$

(3)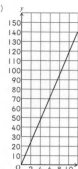

(4) $y = 12 \times \boxed{7} = 84$
   Ans. 84 miles

24 Word Problems with Graphs pp 144, 145

1 (1) **Ans.** $y = 2x$

(2)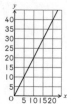

(3) $y = 2 \times 18 = 36$
   Ans. 36 baskets

2 (1) **Ans.** $y = 8x$

(2)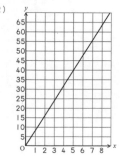

(3) $y = 8 \times 15 = 120$
   Ans. 120 gumballs
(4) $y = 8 \times 9.5 = 76$
   Ans. 76 gumballs

25 Graphs y-Intercept pp 146, 147

1 (1) $\boxed{2}$
(2) $\boxed{-3}$
(3) $\boxed{-1}$
(4) $\boxed{0}$
(5) $\boxed{-3}$

2 (1) $\boxed{1}$
(2) $\boxed{-3}$
(3) $\boxed{-1}$
(4) $\boxed{4}$

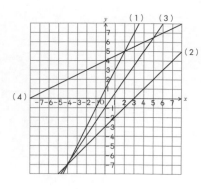

3 (1) $\boxed{-1}$

(2) $\boxed{2}$

(3) $\boxed{-\dfrac{3}{5}}$

(4) $\boxed{\dfrac{16}{7}}$

26 Graphs Slope pp 148, 149

1 (1) $\boxed{2}$

(2) $\boxed{1}$

(3) $\boxed{\dfrac{1}{2}}$

(4) $\boxed{\dfrac{1}{4}}$

2 (1) $\boxed{3}$

(2) $\boxed{1}$

(3) $\boxed{\dfrac{1}{2}}$

(4) $\boxed{\dfrac{2}{3}}$

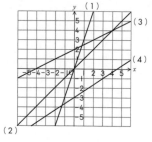

3 (1) $\boxed{2}$

(2) $\boxed{3}$

(3) $\boxed{1}$

(4) $\boxed{\dfrac{5}{2}}$

(5) $\boxed{\dfrac{3}{8}}$

27 Graphs Slope

1 (1) $\boxed{-2}$

x	0	1
y	2	0

(2) $\boxed{-1}$

x	0	1
y	3	2

(3) $\boxed{-\dfrac{1}{2}}$

x	0	1	2
y	-1	$-\dfrac{3}{2}$	-2

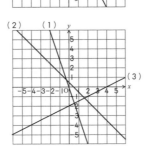

2 (1) $\boxed{0}$

(2) $\boxed{\dfrac{1}{2}}$

(3) $\boxed{-2}$

3 (1) $-3, \ 1$

(2) $-2, \ -6$

(3) $-\dfrac{1}{2}, \ -2$

(4) $-\dfrac{5}{2}, \ 0$

(5) $-6, \ -\dfrac{3}{5}$

28 Graphs

pp 152, 153

1 (1) $3, \ -1$

(2) $1, \ 2$

(3) $-2, \ 3$

(4) $-1, \ -4$

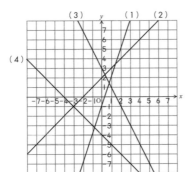

2 (1) $\dfrac{3}{2}, \ 1$

(2) $\dfrac{1}{2}, \ -\dfrac{3}{2}$

(3) $-\dfrac{2}{3}, \ -2$

(4) $-\dfrac{5}{2}, \ \dfrac{1}{2}$

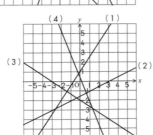

29 Linear Functions

pp 154, 155

1 (1) $\boxed{6}$

(2) $\boxed{-3}$

(3) $\boxed{6}$

(4) $\boxed{6}$

(5) $\boxed{2}$

(6) $\boxed{2}$

(7) $\boxed{6}$

(8) $y = \boxed{2}x + \boxed{6}$

2 (1) $\boxed{2}$

(2) $\boxed{6}$

(3) $\boxed{2}$

(4) $\boxed{-2}, \ \boxed{-\dfrac{1}{3}}$

(5) $\boxed{-\dfrac{1}{3}}, \ \boxed{2}$

(6) $y = \boxed{-\dfrac{1}{3}}x + \boxed{2}$

3 (1) $\boxed{-4}$

(2) $\boxed{-2}$

(3) $\boxed{-4}$

(4) $\boxed{-4}, \ \boxed{-2}$

(5) $\boxed{-2}, \ \boxed{-4}$

(6) $y = \boxed{-2}x - \boxed{4}$

30 Linear Functions

pp 156, 157

1

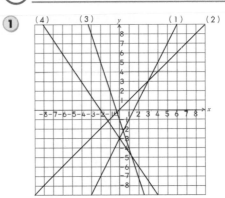

2

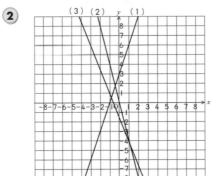

3 (1) $y = \boxed{3}x + \boxed{3}$

(2) $y = -4x$

(3) $y = -\dfrac{5}{2}x - \dfrac{3}{2}$

© Kumon Publishing Co., Ltd.   189

31 Linear Functions

1 (1) $\boxed{2}$
(2) $\boxed{-3}$
(3) $y = \boxed{2}x - \boxed{3}$

2 (1) $y = x + 3$
(2) $y = 3x + 1$

3 (1) $y = -\dfrac{1}{2}x$
(2) $y = 2x + 4$
(3) $y = \dfrac{2}{3}x - 2$
(4) $y = -x + 4$
(5) $y = \dfrac{3}{2}x - 6$
(6) $y = -\dfrac{3}{2}x - 6$

32 Linear Functions

1 (1) $\boxed{3}$, $y = 2x + \boxed{3}$

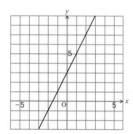

(2) $\boxed{1}$, $y = -\dfrac{1}{2}x + 1$

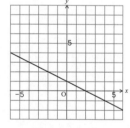

2 (1) $\boxed{1}$, $\boxed{7}$, $y = -3x + \boxed{7}$

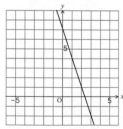

(2) $y = \dfrac{3}{2}x - \dfrac{1}{2}$

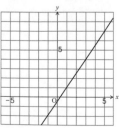

33 Linear Functions

1 (1) $\boxed{3}$,
slope $= \dfrac{5}{\boxed{3}}$

(2) $1 - (-3) = \boxed{4}$,
$4 - (-2) = \boxed{6}$,
slope $= \dfrac{3}{2}$

(3) $4 - 2 = \boxed{2}$, $-3 - 5 = \boxed{-8}$
slope $= -\boxed{4}$

(4) $1 - (-3) = \boxed{4}$,
$-6 - 2 = \boxed{-8}$,
slope $= -2$

2 (1) $\dfrac{9-8}{6-\boxed{2}} = \dfrac{1}{\boxed{4}}$
(2) $\dfrac{7}{4}$
(3) $\dfrac{1}{2}$

(4) $-\dfrac{2}{3}$
(5) -3
(6) $-\dfrac{2}{3}$

34 Linear Functions

1 (1) $m = \dfrac{5 - (\boxed{-1})}{3 - \boxed{0}} = 2$

$y = \boxed{2}x + b$

$\boxed{-1} = \boxed{2} \times 0 + b$

$b = \boxed{-1}$

$y = 2x - 1$

(2)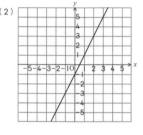

2 (1) $y = 8x + 13$

(2) $y = -\dfrac{2}{3}x + \dfrac{5}{3}$

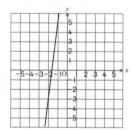
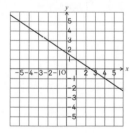

35 Linear Functions

1 (1) Ans. $y = -x + 7$
(2) Ans. $y = \dfrac{1}{2}x + 2$
(3) Ans. $y = -\dfrac{5}{3}x$

2 (1) Ans. $y = -x + 7$
(2) Ans. $y = \dfrac{1}{2}x - 1$
(3) Ans. $y = -5x$
(4) Ans. $\boxed{0}$, $y = 6x + 2$

36 Linear Functions

pp 168,169

1 (1) $y = \boxed{-2x} + 5$

$\boxed{-2}$

$\boxed{5}$

(2) $\boxed{3}, \boxed{-4}$

(3) $\boxed{-\dfrac{1}{2}}, \boxed{-1}$

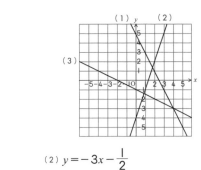

2 (1) $3y = \boxed{-2x} + 12$

$y = \boxed{-\dfrac{2}{3}}x + 4$

$\boxed{-\dfrac{2}{3}}$

$\boxed{4}$

(2) $y = -3x - \dfrac{1}{2}$

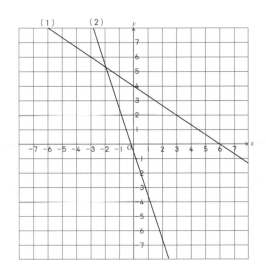

37 Simultaneous Linear Equations

pp 170,171

1 (1) $y = \boxed{-3x} - 1$

$y = x + 3$

$(x, y) = (-1, 2)$

(2) $\begin{cases} y = -3x - 1 \\ y = x + 3 \end{cases}$

$(x, y) = (-1, 2)$

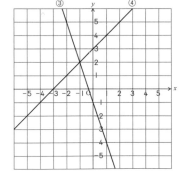

2 (1) $(1, -3)$

(2) $(1, -3)$

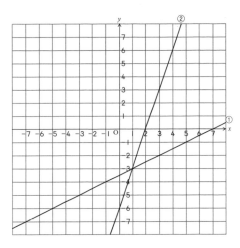

38 Horizontal and Vertical Lines

pp 172,173

1 (1) $y = -3$

(2) $y = \boxed{4}$

(3) $y = \dfrac{3}{2}$

(4) $y = -\dfrac{1}{2}$

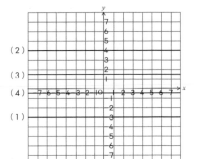

2

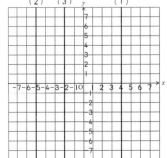

3

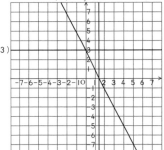

(39) Parallel Lines
pp 174,175

1

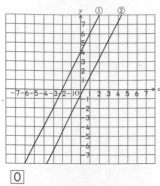

0

2

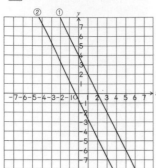

3 (1) $\boxed{2}$, $\boxed{2}$, $y=\boxed{2}x+b$ **Ans.** $y=2x-4$

(2) **Ans.** $y=-\dfrac{1}{4}x-1$

(40) Perpendicular Lines
pp 176,177

1

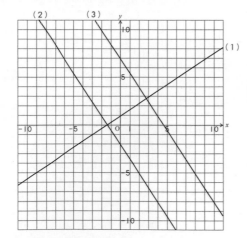

2 (1) $\boxed{-\dfrac{4}{3}}$

(2) $\boxed{\dfrac{2}{7}}$

(3) $\boxed{-\dfrac{1}{6}}$

3 (1) $\boxed{-\dfrac{3}{2}}$, $\boxed{-\dfrac{3}{2}}$, $y=\boxed{-\dfrac{3}{2}}x+b$ **Ans.** $y=-\dfrac{3}{2}x-7$

(2) **Ans.** $y=2x+5$

(41) Perpendicular Lines
pp 178,179

1 (1) **Ans.** $y=-\dfrac{2}{3}x-3$

(2) **Ans.** $y=\dfrac{1}{10}x+\dfrac{19}{40}$

2 (1) **Ans.** $y=\dfrac{3}{2}x+\dfrac{11}{2}$

(2) **Ans.** $y=-\dfrac{2}{3}x-1$

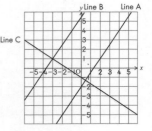

(42) Review
pp 180,181

1 (1) **Ans.** $(x,\ y)=(-2,\ 4)$, **Ans.** $(x,\ y)=(-2,\ 4)$

2 Let x be the number of people in Nori's family, and y be the number of rabbits. There are $3y$ chicks.

$$\begin{cases} 2x+3=y \\ 3x+27=3y \end{cases}$$

Ans. 6 people in Nori's family
45 chicks

3 (1) $x<-\dfrac{16}{3}$ (3) $x<5$

(2) $x\le-\dfrac{2}{5}$ (4) $x\ge\dfrac{16}{11}$

4 (1) **Ans.** No solution   (2) **Ans.** $x\le\dfrac{2}{13}$

(43) Review
pp 182,183

1

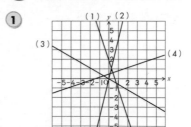

2 (1) **Ans.** $(-1,\ 3)$

(2) **Ans.** $(-1,\ 3)$

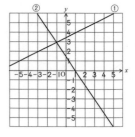